Lean Karlo Santos Tolentino
Editor

Aquaponics

Sustainable Farming, Ecology and Innovation

DOI: https://doi.org/10.52305/SYHM3238

NOTICE TO THE READER

Library of Congress Cataloging-in-Publication Data

ISBN: 979-8-89530-464-8 (Softcover)
ISBN: 979-8-89530-584-3 (eBook)

Published by Nova Science Publishers, Inc. † New York

Contents

Preface

In an era where sustainable agriculture and environmental stewardship are becoming increasingly vital, this book explores innovative approaches to integrating technology with traditional farming practices. The chapters presented delve into the synergy between aquaponics, hydroponics, and cutting-edge IoT-based systems, demonstrating how these technologies can address critical challenges in food production and environmental management.

Chapter 1, "The Critical Role of Water pH in Aquaponics Systems," emphasizes the delicate balance required in aquaponics, where the interdependence of fish, plants, and bacteria hinges on maintaining optimal water pH levels. This foundational understanding sets the stage for successful aquaponic systems that maximize productivity without compromising the health of any component.

Chapter 2, "A Fully Automated Positioning and Handling System for Indoor Hydroponic Crop Harvesting," introduces a fully automated harvester designed to streamline the indoor hydroponics process. By integrating robotics and precise motor control, this system significantly reduces human labor while enhancing the efficiency and consistency of crop harvesting, a crucial step towards modernizing agriculture.

Chapter 3, "CYNOSENS: An IoT-Based Harmful Algal Bloom Prediction System Using Random Forest Regression Through Water Quality Parameters," addresses a pressing issue in aquaculture: harmful algal blooms. This chapter presents an IoT-enabled monitoring system that predicts the onset of harmful blooms with high accuracy, offering a proactive tool for protecting aquatic environments and the livelihoods that depend on them.

Finally, Chapter 4, "SMAQ: A Smart Aquaponics System Using IoT Technology," explores the application of smart technology in aquaponics to improve system sustainability and productivity. By automating the monitoring and control of key environmental parameters, this system offers a model for urban agriculture that meets the demands of growing populations while minimizing environmental impact.

Together, these chapters provide a comprehensive overview of how integrating technology with traditional agricultural practices can lead to more efficient, sustainable, and resilient food production systems. Whether you're a researcher, practitioner, or enthusiast in the fields of agriculture and technology, this book offers valuable insights into the future of farming.

We extend our heartfelt gratitude to the contributing authors for their dedication and perseverance in completing the chapters despite various challenges. We also appreciate the diligent efforts of the editors at Nova Science Publishers.

Lean Karlo Santos Tolentino,
Associate Professor,
Dean, College of Engineering
Technological University of the Philippines, The Philippines

About the Editor

Lean Karlo Santos Tolentino

College of Engineering, Technological University of the Philippines, Manila, Philippines

Associate Professor Dr. Lean Karlo Santos Tolentino received a B.S. degree (cum laude) in electronics and communications engineering from the Technological University of the Philippines (TUP), Manila, in 2010, the M.S. degree in electronics engineering with a specialization in microelectronics from Mapúa University, in 2015, and a Ph.D. degree in electrical engineering (system-on-chip group) from National Sun Yat-Sen University, Kaohsiung, Taiwan, in 2024.

From 2010 to 2013, he was an IC Layout Engineer with the Mask Design Division, ROHM LSI Design Philippines, Inc., Pasig, Metro Manila, Philippines. Since 2015, he has been an Associate Professor with TUP. From 2017 to 2019, he was the Head (Chair) of the Department of Electronics Engineering, TUP. From 2019 to 2020, he was the Director of the Office of the University Extension Services (UES), TUP. As of March 2024, he has been the Dean of the College of Engineering, TUP. His research interests include artificial intelligence, IC design, and power electronics.

Dr. Tolentino has been a member of the Institute of Electronics Engineers of the Philippines (IECEP) since 2011, where he has been the Director of the Manila Chapter, since 2022. Since 2017, he has been a member of the Technical Committee on Audio, Video, and Multimedia Equipment (TC 59), the Technical Committee on Electromagnetic Compatibility (TC 74), and the Technical Committee on Information Technology/Subcommittee on Artificial Intelligence and Quantum Technologies (TC 60/SC 4) of the Bureau of Philippine Standards, which operates under the Department of Trade and Industry of Philippine Government. In 2019, he was elected as a Regular Member of the National Research Council of the Philippines (NRCP). From 2022 to 2023, he was the elected President of the Association of Filipino Scholars in Taiwan (AFST). Since 2022, he has been a member of the Subcommittee for General Engineering and Applied Sciences (GEAS), Philippines. This subcommittee is responsible for assisting in the development of the syllabi and tables of specifications (TOS) for the Electronics Engineers Licensure Examination (EELE) and the Electronics Technicians Licensure Examination (ETLE). He holds a valid license as a Professional Electronics Engineer (PECE) in Philippines and was conferred as ASEAN Engineer. Since 2024, he has been the Interim Chair of the IECEP

Microelectronics Society.

E-mail: lean_dll@yahoo.com; leankarlo_tolentino@tup.edu.ph

Chapter 1

The Critical Role of Water pH in Aquaponics Systems

Mostafa A. M. Soliman[1,*], PhD
Safwat A. A. Gomha[2], PhD
and Mohamed A. A. Abo-Taleb[3], PhD

[1]Utilization by-Products Department, Animal Production Research Institute (APRI), Agriculture Research Center, Ministry of Agriculture, Dokky-Giza, Egypt
[2]Central Laboratory of Aquaculture Research (CLAR)-Agriculture Research Center, Ministry of Agriculture), Abo-Hammad-El-Sharkia Gov, Egypt
[3]Regional Center for Food and Feed, Agriculture Research Center, Ministry of Agriculture, Giza, Egypt

Abstract

Integration between fish and plants in the aquaponic system is no longer an idea to benefit from fish waste to grow plants depending on bacterial activity for the oxidation of ammonia to nitrate. However, there is a distinct relationship between the three beings that rely on coordination between them through an in-depth study of this relationship. The success of this relationship depends on a good study of water quality parameters, which greatly affect the various organisms in the aquaponic system. The most important factor that significantly affects the success of the aquaponic system is pH. This is due to the difference in the appropriate pH range for each organism on which the aquaponic system depends. Therefore, they must be coordinated in the pH range that helps the production process succeed without affecting them.

* Corresponding Author's Email: mostafa.soliman242@gmail.com

In: Aquaponics: Sustainable Farming, Ecology and Innovation
Editor: Lean Karlo Santos Tolentino
ISBN: 979-8-89530-464-8

Keywords: aquaponics system, water quality, pH, fish, plant

Introduction

The development of agro-ecosystems with an emphasis on crop production is focused on increasing productivity while minimizing environmental effects. In order to integrate production systems, the combination of fish production systems and crop production demonstrated promise (Gooley & Gavine, 2003; Lenz et al., 2021). According to Prapti et al. (2022), heat degree (20 percent), dissolved oxygen (18 percent), and pH (17 percent) are the most highly prioritized water functionality factors taken into account in IoT-based aquaculture. pH is ranked as the third most important metric in this review analysis. The pH scale is utilized to determine the amount of hydrogen ions present in water. Any pH variation makes low-growing and low-producing bacteria more susceptible to infections and illness. Generally speaking, fish die when the pH value is less than 4.5 and greater than 10. The pH is significant because it also affects the ratio of ammonium to ammonia in water. Moreover, in open systems, pH fluctuates throughout the day due to biological processes, including photosynthesis and respiration (Praptiet al., 2022).

The pH is one of the factors that affect the different properties of water. Therefore, it affects organisms that live in the aquatic environment, whether microscopic, such as phytoplankton and zooplankton, or aquatic plants and fish. Where, Sørensen (1909) defined pH according to the number of hydrogen ions (using modern terminology) as pH = lg(cH/c°), where cH represents the hydrogen ion concentration in mol dm-3. and c° = 1 mol dm-3 is the typical proportion. As a result, Sørensen and Linderstrøm-Lang (1924) decided that describing pH regarding the corresponding activity of hydrogen ions in solution pH is more appropriate.

$$pH = -\lg aH = -\lg(mH\gamma H/m^\circ) \quad (1)$$

In this equation, aH represents the activity ratio (molality basis), γH is the molal movement coefficient of the hydrogen ion H+ at mH, and m° is the typical molality. The purpose of pH is to quantify the activity of hydrogen ions in a medium. While given that it is described in regards to an amount that a thermodynamically valid technique cannot determine, equation 1 may only be considered as a theoretical description of pH.

As the hydrogen ion is involved in many reactions in water, its activity has a significant impact on the success or failure of some systems available in fish farming, the most important of which is aquaponics and bio floc systems. Aquaponics combines hydroponic nourishment delivery with fish farming, using nutritional-rich effluent from fish farming operations as a fertilizer solution for horticultural vegetables and fruits. Since the components of fish feces supply nourishment, aquaponic systems often have lower quantities of certain nutrients, especially potassium, calcium, magnesium, and iron (Rakocy et al., 2004).

An aquaponics system's nutrient solution serves as more than just a passive conduit enabling the transfer of fish-to-plant fertilizers. Due to the high concentration of dissolved ions and organic compounds produced by metabolic processes in fish and feed elimination, the chemical composition of aquaponics feeding solutions is complicated. The chemistry of the substance of aquaponics fertilizer fluids can be affected by the interaction of the main ions. The pH of the fluid could have a substantial influence on the uptake of nutrients by vegetable roots (White, 2012). Note that Aquaponic fertilizer solutions are more challenging to maintain than hydroponic nutrient solutions. Numerous variables influence the aquaponic solution, including fish consumption level, hydrodynamic loading rate, and pH (Wongkiew et al., 2017; Chu et al., 2023).

By merging fish and plant systems, the pH is an important aspect in managing the nutrition level, and it must be managed for three distinct organisms concurrently (Wongkiew et al., 2017). Unlike traditional soil cultivation, aquaponic plant development uses nutrients dispersed into liquid instead of deposited onto the outermost layer of soil granules. Since aquaponics is associated with the cultivation of aquatic life in linked systems, the pH of the water used for irrigation is frequently maintained at a level that encourages fish productivity and nitrification over plant production, with both ecosystems balancing at a pH level of 7.0 (Tyson et al., 2007). The pH ranges of 6.0-6.5 were suitable for the growth of vegetables but somewhat below those for fish rearing (pH 6.5-9.0), while the development of nitrifying microbes (pH 7.0-8.0), according to Li et al. (2019). The aquaculture and horticulture components are separated in more recent aquaponic system designs. These decoupled systems enable independent unit optimization (Blanchard et al., 2020).

Furthermore, aquaponics can be described as a sustainable system, considering the pH of the aquaponics solution. Moreover, the decoupled framework will improve the usage of fish sludge by adjusting the pH.

Water Parameter under Aquaponics System

Controlling water integrity is vital for enhancing the health and growth of aquatic animals and plants. In a RAS/aquaponic system, the most important variables to monitor are the influence of bacteria on waste products containing nitrogen, pH, alkalinity, oxygen dispersion, temperature, carbon dioxide, and Semi-thawed particles (Timmons & Ebeling, 2007).

pH Effect on Fish

If production protocols are followed, pH is the independent parameter that needs to be measured frequently (Rakocy et al., 2003). Exposure to aquatic organisms to pH extremes is distressing or deadly; however, in aquaculture, the secondary impacts of pH and their combinations with additional factors are frequently of greater significance than the obvious adverse impacts (Doudoroff, 1956). It was discovered that the pH fluctuations in the liquid medium had an impact on fish growth, reproduction, illness susceptibility, and the formation of embryos (Doudoroff, 1956; Kwain, 1975). Multiple investigations indicate that increased pH values have a deleterious influence on aquatic organisms' growth and reproduction, resulting in widespread mortality in aquatic cultures (Jezierska & Witeska, 1995; Zweig et al., 1999; Zaniboni-Filho et al., 2002; Scott et al., 2005; Zaniboni-Filho et al., 2009; Nchedo & Chijioke, 2012), being shown in Table 1. As stated by Zaniboni-Filho et al. (2002), most freshwater aquatic organisms require a pH between 6.5 and 9.0 for optimal growth and to stay healthy. The mixed catfish (*Heterobranchusbidorsalis* x *Clarias gariepinus*) raised at pH 7.0 and 7.5 had considerably faster specific growth rates and feed conversion ratios than the fish kept at pH 6.0 and 8.0, according to Njoku et al. (2007). Increasing the pH offers a positive effect on Nile tilapia (O. niloticus) development; however, this does not apply to other kinds of fish (Iqbal et al., 2012). Fish species in neotropical environments favor pH ranges of 6 to 8 (Lopes et al., 2001; Townsend & Baldisserotto, 2001; Baumgartner et al., 2008).

Furthermore, water pH affects several metabolic processes, and whenever aquatic organisms are subjected to excessively acidic or highly alkaline fluid, their gills' ionic equilibrium deteriorates, eventually leading to significant death. (Lloyd and Jordan, 1964; Alabaster and Lloyd, 1980; Freda and Mcdonald, 1988; Mcgeer and Eddy, 1998). The developmental capability of most aquatic organisms is severely hampered by pH values below 6.0 or over 9.0 (Parra & Baldisserotto, 2007). Every pH shift, either upward or downward from the optimal range, may have an adverse effect on the physiological or metabolic processes of fish, including growth, behavior during reproduction, and dispersal across the ecosystem (Boyd, 1998; Zweig et al., 1999), illustrated in Figure 1. In fish, an acidic environment has either a negative or no effect on proliferation (Mount, 1973; Leivestad et al., 1976; Menendez, 1976; Jacobsen, 1977). Reduced growth has been recorded in different fishes at an acidic pH of 5.5 (Menendez, 1976; Craig & Baksi, 1977; Ndubuisi et al., 2015). Because low pH reduces the activity of phagocytes of channel catfish neutrophils, it affects the innate immune response (Ainsworth et al., 1991). The ideal pH range for silver catfish larvae to thrive and develop is 8.0-8.5 (Lopes et al., 2001), while juvenile development is reduced at pH 5.5 or 9.0 as opposed to pH 7.5 (Copatti et al., 2005). Acidic conditions cause fish to lose their ability to maintain homeostasis, which increases the excretion of H^+ and NH_4^+ in their urine (Wood, 2001; Bolner & Baldisserotto, 2007). One theory for fish mortality in very acidic conditions is that they cannot regulate their inside ion concentrations due to a drop in ion absorption rates (Laurent et al., 2000).

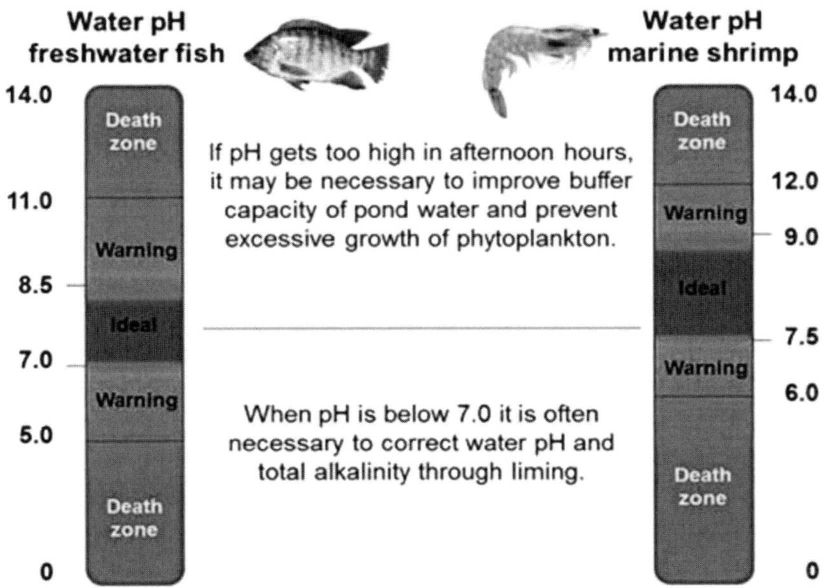

Figure 1. pH monitoring guidelines for farmed freshwater fish and marine shrimp.

Table 1. Impact of pH on fish and other aquatic organisms

pH	Effects
4	Acid-death limit
4-5	No reproduction[1]
4-6.5	Numerous kinds are growing slowly[1]
6.5-9	Ideal region
9-11	Slow growth and adverse reproductive effects
11	Alakaline death point

[1]Certain species of fish in streams coming from rainforests thrive at low pH (Boyd, 2000).

pH Effect on Nitrification Bacteria

Nitrification, which transforms harmful NH_3 into NO_3^- for plant uptake, is an important phase in aquaponics (Wongkiew et al., 2017; Wongkiew et al., 2018; Tan et al., 2022; Chu et al., 2023). Nitrifying bacteria constitute perhaps one of the most essential species in any aquaponic environment, according to (Yep & Zheng, 2019). This is due to the inorganic ammonium is the fish's main excretory nitrogen product and, even at low concentrations, affects their growth (Benli & Köksal, 2005).

The bacterial ecosystem's importance in aquaponics is one of its characteristics (Kasozi et al., 2021). Goddek et al. (2015) summarized the findings of earlier research showing that the three primary nitrification organisms in aquaponic systems are *Nitrobacter*, *Nitrosomonas*, and *Nitrospira*. However, they rapidly shifted the belief that Nitrosomonas is the main bacteria that oxidizes nitrite and that Nitrobacter is the main bacteria that oxidizes ammonia. An examination of the populations of bacteria in various parts of an aquaponic system published by Schmautz et al. (2017) found that *Nitrospira* comprised 3.9% of the biofilter microbial population, while *Nitrobacter* and *Nitrosomonas* only constitute 0.11% and 0.64%, respectively. Based on thermodynamics and the movement of electrons, a balanced nitrogen-fixing system should have a ratio of ammonia-oxidizing bacteria (AOB) to nitrite-oxidizing bacteria (NOB) of 2:1 (Hooper et al., 1997). However, other investigators have also reported controversial is proportionate AOB/NOB ratio (Mari et al., 2012 and Ramdhani et al., 2013).

Lately, depending on the biochemistry involved, the procedure for nitrification is carried out by three distinct types of microorganisms (Daims et al., 2015; Stein & Klotz, 2016; Holmes et al., 2019). Microorganisms in the first group convert NH_4^+ and NH_3 to NO_2^- by nitridation. The following group of bacteria transforms NO_2^- to NO_3^- by nitrification processing, whereas the last group (Third) changes NH_4^+ to NO_3 (Comammox)—the initial step in nitrification, including bacteria from the taxa (Figure 2). Nitrosomonas, Nitrosococcus, Nitrosospira, Nitrosolobus and Nitrosovibrio. Nitrobacter, Nitrococcus, Nitrospira, and Nitrospina are members of group two (Hagopian & Riley, 1998). According to Daims et al. (2015), group three is from genealogy II of Nitrospira. These bacteria are chemoautotrophs that employ oxygen as the last recipient of electrons and NH_4^+ as a form of energy to fix CO_2 (Hagopian & Riley, 1998).

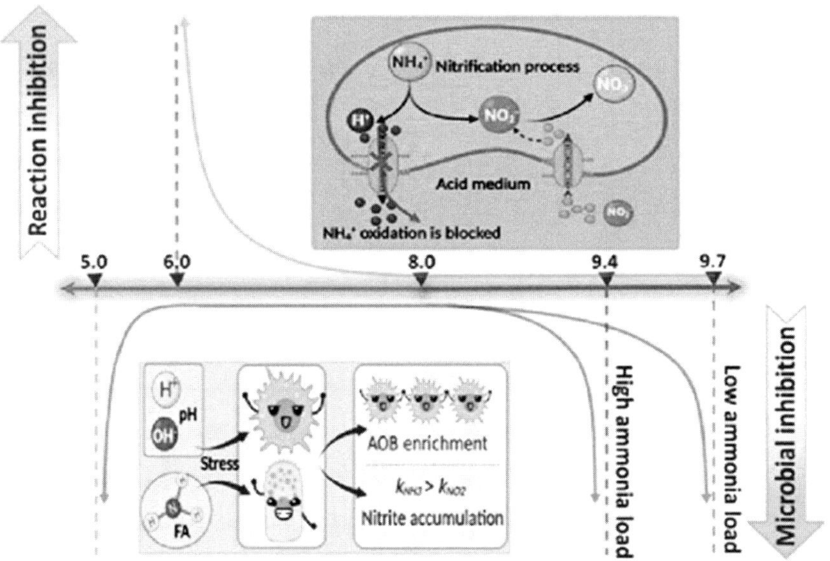

Figure 2. Both mechanisms have been combined to produce a comprehensive picture of the influence of pH stress on nitrification, enhancing our knowledge of nitrifier adaptation and nitridation control in wastewater treatment processes (Yue et al., 2023).

Nitrifying microorganisms may develop globally in the system; however, they are normally more concentrated in bio-filters if they are present. The perfect conditions for nitric oxide are between 25 and 30°C with a pH of 7 and 9 (optimally 7.8 according to Antoniou et al. (1990)). Also, the oxygen concentration was less than 20 mg per liter (Rakocy et al., 2006). Temperatures between 15 and 20°C are ideal for the *Nitrosomonas* spp. bacteria, which turns ammonia into nitrite. The bacteria that transform nitrite to nitrate, *Nitrobacter* spp., however, operate inadequately at those heat conditions and favor a pH range of 7.5-8.6 with at least 1 mg L^{-1} oxygen present in the medium (Huey et al., 1984 and Chen and Cheng 2000).

Table 2. Optimal pH range of major microorganisms in aquaponics systems and microbiological processes

The various processes by bacteria		Genera	Reference	pH	Reference
Nitrification	Ammonia oxidization	*Nitrosomonas, Nitrosococcus, Nitrosospira, Nitrosolobus, Nitrosovibrio*	Ebeling et al., (2006); Rurangwa and Verdegem (2015); Schreier et al., (2010); Schmautz et al., (2017)	7.6 3.5-5.5 4.5-7.6 6.5-7	Derikvand et al., (2021); Gieseke et al., (2006); Fumasoli et al., (2017); Zhou et al., (2021)
	Ammonia oxidization by archaea	*Trosopumilus, Nitrososphaeras*	Bartelme et al., (2019)	----- 7.5	Stieglmeier et al., (2014)
	Nitrite oxidation	*Nitrobacter, Nitrospira, Nitrococcus, Nitrospina*	Ebeling et al., (2006); Schmautz et al., (2017); Wongkiew et al., (2018)	7.2-8.2 4.5-7.6 7-9 7-9	Kasozi et al., (2021); Derikvand et al., (2021); Gieseke et al., (2006)
	Complete ammonia oxidation	*Nitrospira*	Daims et al. (2015); Bartelme et al. (2019)	4.5-7.6	Derikvand et al., (2021)
Denitrification		*Dokdonella, Thermomonas*	Schmautz et al., (2017)	7.6-9 -------	Derikvand et al., (2021)
Mineralization		*Pseudomonas, Flavobacterium, Sphingobacterium, Arcobacter*	Eck et al., (2019); Wongkiew et al., (2018)	7-9 7.2-7.4 7.2 5.5-8	Zhou et al. (2021). Bernardet and Bowman (2006); Song et al. (2021); D'sa and Harrison (2005)
Anaerobic ammonium oxidation (Anammox)		*Brocadia*	Schmautz et al., (2017)	7.6–7.8	Okabe et al., (2021)
Sulfate reduction		*Fusibacter, Bacteroides, Desulfovibrio, Dethiosulfovibrio*	Schreier et al., (2010); Somerville et al., (2014)	5.2-8.3(7.2) 6-7 6.6-8.2 (7.8) 6-8 (6.5)	Brioukhanov et al., (2023); Fultz et al., (2021); Alazard et al., (2003); Grabowski et al., (2022)

The various processes by bacteria		Genera	Reference	pH	Reference
Organic phosphorus mineralization		*Modestobacter*	Kasozi et al., (2020)	6.0–11.0,	Jiang et al., (2021)
Iron cycling		*Acidibacter*	Kasozi et al., (2020)	2.5-4.5	Falagán and Johnson (2014)
Nitrogen fixation		*Pontibacter, Pseudonocardia*	Kasozi et al., (2020)	6.5-7.5 7	Chhetri and Seo, (2021); Kiranmayi et al., (2011)

Nitrosomonas and *Nitrobacter* are the most frequent nitrifying bacteria in aquaponics, with pH resistance values varying from 7.2-7.8 and 7.2-8.2, respectively (Somerville et al., 2014; Sallenave, 2016). In order to encourage nitrification in the bio-filters, which enables more of the poisonous NH_3 to be transformed to NO_3, The aqua farming industry keeps pH values around 7.0 and 8.5 (Timmons et al., 2001). At the same time, the ideal pH for ammonia-oxidizing bacteria (AOB) and nitrite-oxidizing bacteria (NOB) is 7.2 to 8.2, while pHs 5.8 and 6.5, respectively, limit their growth (Gieseke et al., 2006; Zhang et al., 2019; Tan et al., 2022). Where low pH has been demonstrated to change the microbial populations in an aquaponic system (Wongkiew et al., 2018).

Generally, each component's percentage in TAN equilibrium varies according to the pH and warmth of the water. At 28°C, the fraction of NH3 rises by about ten, reaching 0.2%, 2%, and 18% of the TAN at pH 6.5, 7.5, and 8.5, respectively (Francis-Floyd et al., 2009). During liquid pure culture, activity frequently terminates below the pH of 5.5 (Gieseke et al., 2006; Zhang et al., 2019; Tan et al., 2022).

Nitrification activity in fish farming biological filters appears to be most effective between pH 7.0 and 9.0 (Zhang et al., 2019; Tan et al., 2022). Wongkiew et al. (2017) illustrated that nitrification is reduced under pH 6.0. However, the impact of pH on the population of bacteria in aquaponics systems is unclear. For example, Wongkiew et al. (2018) observed that Fusobacteria predominated over Acidobacteria in biofilters at pH 6.8–7.0, but this abundance varied in plant roots at the same pH. However, in the root samples, where Proteobacteria predominated, the relative abundance of *Acidobacteria* dropped at pH 5.2. Since it cannot be feasible to reach an ideal pH for fish, bacteria, and plants in aquaponics systems, keeping pH values between 6.5- 7.0 achieves a balance between nitrification and nutrient availability (Somerville et al., 2014; Sallenave, 2016; Wang et al., 2023), the appropriate pH ranges for various bacterial species are displayed in Table 2.

pH Effect on Minerals and Elements

There are two key benefits of aquaponics for nutrient cycling. First, the use of hydroponic production in association with a recirculating aquaculture system prevents the disposal of farming solid waste that is high in incorporated nitrogen and phosphorus into existing polluted groundwater (Buzby & Lin, 2014; Guangzhi, 2001; van Rijn, 2013), secondly, enables it possible to fertilize crops without soil with an organic solution (Goddek et al., 2015; Schneider et al., 2004; Yogev et al., 2016) rather than mineral fertilizers derived from decreasing natural resources (Schmautz et al., 2016). The fish feed and the aqueous medium that enters the system (containing Mg, Ca, and S) are the main sources of nutrients in an aquaponic system (Delaide et al., 2017; Schmautz et al., 2016). According to Timmons and Ebeling (2013), a traditional fish diet comprises 6–8 macro components. This comprises 6-8% organic nitrogen, 1.2% organic phosphorus, and 40–45% organic carbon, as well as about 25% protein for herbivorous or omnivore fish and about 55% protein for carnivorous fish (Boyd, 2015). Lipids can also be derived from plants or fish (Boyd, 2015). The four most typical necessary elements for plants, Ca, K, Mg, and Fe, are limiting in the solution in aquaponic systems (Rakocy, 2003; Villarroel et al., 2011). Considering fish have low prerequisites for numerous minerals (Fe, Mn, Mg, Cu) and even lower requirements for K (only 1% of the composition) (Seawright et al., 1998). As a result, there are fewer of these elements in aquaculture wastewater and fish feed (IAFFD, 2018; Rakocy, 2003). Fish meal is the main component of the majority of fish feeds (Cerozi and Fitzsimmons, 2017a,b). Although fish meal is rich in phosphorous and amino acids, which are organic forms of nitrogen, it is deficient in K and several micronutrients that plants need, including Fe, Mn, and Copper (IAFFD, 2018; Savidov et al., 2007).

Considering pH is a logarithmic function, a modification of one unit in pH corresponds to a tenfold alteration in H+ concentration. As a result, every unit change in pH can significantly impact ion accessibility for crops. For the best possible nutrient absorption, most plants need a pH range between 6.0 and 7.0 (Resh, 2013). Because of systemic chemical interactions and an inadequate pH of the system's water, these nutrients are further rendered inaccessible to plant absorption (Yep & Zheng, 2019).

The pH range for plant beds is 5.5-5.8 (Bugbee, 2004). The accessible P for vegetables is affected by pH (Asao, 2012). Most P changes to insoluble complexes as pH rises over 7.0; in addition, 30-65% of the P is left in solid fish sludge, which is unobtainable to crops (Asao, 2012). The forms of phosphorus in liquid alter as a perform of pH. The pKs for converting H3PO4 to H2PO4- and then to HPO4-2 are 2.1 and 7.2, respectively (Schachtman et al., 1998). As a result, phosphorus primarily exists in the form H2PO4- in the pH range aquaponics systems maintain, while H3PO4 and HPO4-2 possess less activity (Cerozi & Fitzsimmons, 2016). Additionally, when pH is higher than 6.5, plants find it difficult to take in Fe, Cu, zinc (Zn), boron (B), and Mn (Asao, 2012).

The overall concentration of orthophosphate decreased as the pH rose. However, it was not until pH 8.5 that there was a noticeable departure from the pH range (pH 3.0-5.5) where orthophosphate was most readily available. However, only pH 10.0 significantly altered orthophosphate concentration compared to the normal pH of fertilizer solutions used in aquaponics (pH 6.8) (Cerozi & Fitzsimmons, 2016). Minerals dissolve at varying frequencies and do not build up consistently (Seawright et al., 1998; Rakocy & Hargreaves, 1993). This influences the amounts in liquid. All of the microorganisms involved, as well as the chemistry and physical processes of solubility, are poorly known (Van Rijn, 2013; Krom et al., 2014).

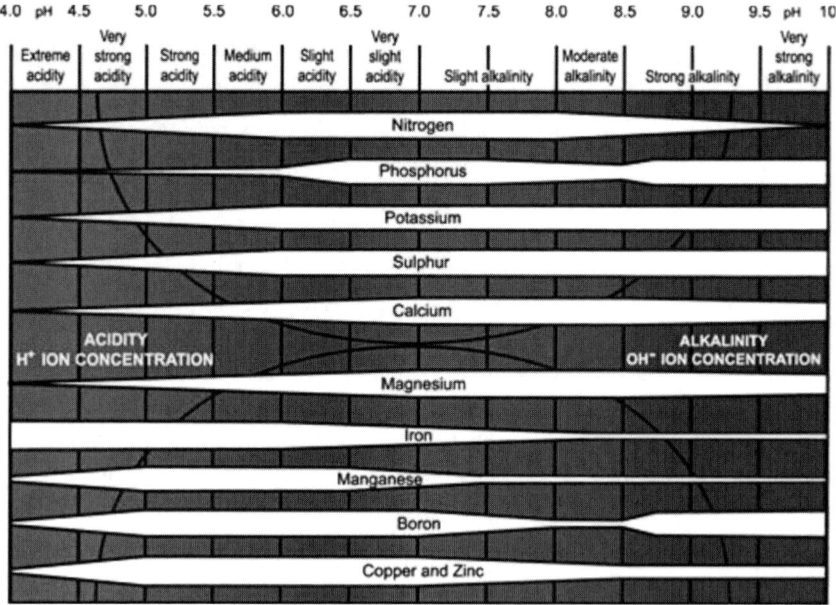

Figure 3. The dimension of the white bar reflects the accessibility of various nutrients at various pH bands; the larger the bar, the more obtainable the nutrient (redrawn for PDA from Truog, (1946).

A pH of 5.8 or lower maintains the majority of ions in a solution, whereas a pH of >6.5 can result in nutrient shortage caused by nutrient deposition and depletion (Bugbee, 2004). The availability of iron, manganese, and zinc decreases when pH rises from 6.5 to 7.5 or 8.0. On the other hand, the availability of phosphorus and molybdenum is impacted in the opposite manner, increasing with increasing pH levels. Bicarbonate ions (HCO_3^-) might be present in high enough concentrations at very high pH levels to obstruct other ions' normal uptake, which is harmful to optimal growth(Resh, 2013). The aquaculture and horticulture units are decoupled in newer aquaponic system designs. Each unit in these decoupled systems can be optimized independently (Blanchard et al., 2020). Elevated pH levels (over 7.0) result in the precipitation of insoluble and inaccessible salts such as Fe_2^+, Mn_2^+, PO_4^{3-}, Ca_2^+, and Mg_2^+ (Resh, 2013), as demonstrated in Figure 3. Therefore, it can be described as a sustainable element of the aquaponics system, considering the pH of the aquaponics water.

pH Effect on Plants

Overall, because they thrive in water that is high in nitrogen, Get a short growing period and needless nourishment, and are highly sought-after all over the world, leafy greens are the favored plants that grow in integrated aquaponics systems (Bailey & Ferrarezi, 2017). Compared to leafy vegetables, flowering crops are more valuable economically, but producing them in aquaponic systems is more challenging because of their higher phosphorus and potassium requirements, greater vulnerability to diseases and insect pests, and lengthier growth periods (Rakocy, 2003). As a result, considering the proper pH level for the type of plants and achieving a balance with the appropriate pH level for the elements in their available form will aid in the success of the aquaponic system (Figure 4).

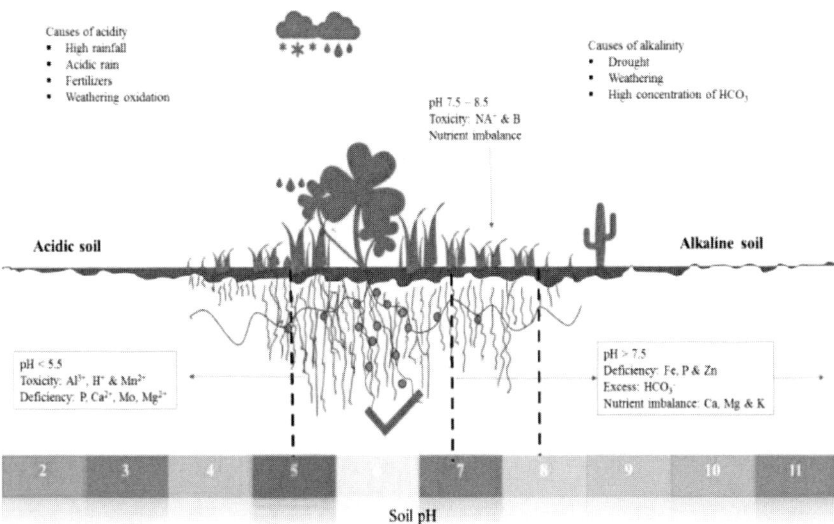

Figure 4. A comparison of different soils in terms of pH, accessibility to nutrients, shortcomings, and instabilities (Msimbira & Smith, 2020).

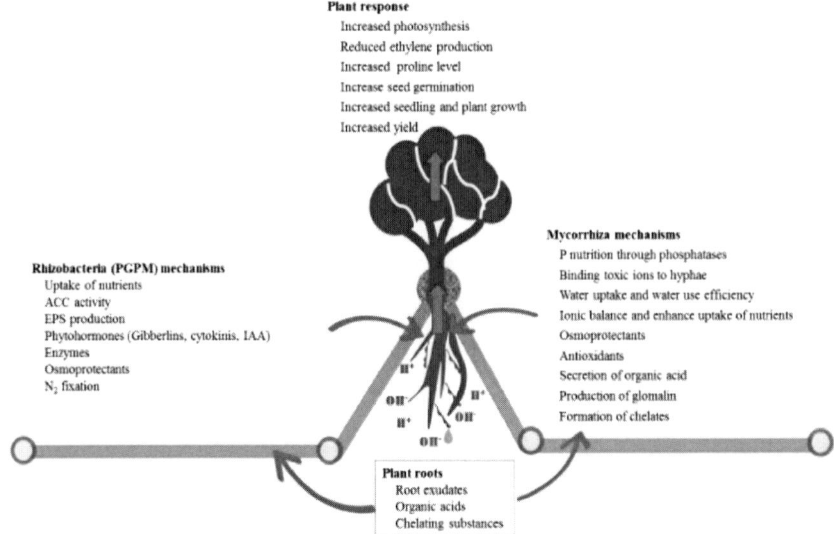

Figure 5. Distinctions in the methods employed by bacteria and fungi to lessen different environmental pressures (Msimbira & Smith, 2020).

While acidity pH values might be regarded as a challenge, Rakocy (2003) noted that if the pH in a developed system remains constant, it could have been due to extensive denitrification. Excessive denitrification reduces nitrogen utilization efficiency (NUE) and the availability of nitrogen for plant consumption (Figure 5). Aquaponic plant growth utilizes nutrients that are dissolved in solution as opposed to those that are trapped in the outermost layer of soil grains. Since hydroponics is related to aquaculture production in coupled systems, the pH of irrigation water is frequently maintained at a level that promotes fish production and nitrification rather than plant cultivation, synchronizing the two processes at a pH of 7.0 (Tyson et al., 2007). In addition, for plants to develop and produce effectively, survive a long time, and be adequately taken via the roots of plants, nutrient solutions need to have a neutral pH of 6.0 to 7.5. In an alkaline condition, micronutrients like Mn, Zn, and Fe become chemically bound and are insufficiently assimilated by plant roots, leading to nutritional shortages and restricted growth (Sahubawa et al., 2020).

The solution for aquaponic systems pH is equilibrium among the demands of vegetables and microorganisms. Microbial nitrification of ammonia to nitrite and nitrite to nitrate is greatest around pH 8.5, even though plant absorbance of nutrients for a variety of crops is excellent near pH 6.0.; as a result, pH in integrated systems is maintained close to pH 7.0 (Wortman, 2015). According to a new research report, pH 6.0 was adept for plant development and efficiency in using nitrogen in aquaponics systems at the expense of elevated N_2O releases due to considerable denitrification (Zou et al., 2016). Plants are capable of absorbing the liberated orthophosphate ions $H_2PO_4^{2-}$ and HPO_4^{2-}. (Becquer et al., 2014). In the pH spectrum of 5.6-8.5, the amount of phosphate intake reduces as the pH of the external solution rises, and this seems to be caused by a reduction in the quantity of H_2PO_4, which is the substance of the proton-coupled phosphorus symporter in the plasma membrane; however, a lessen in pH may improve the efficiency of proton-coupled solute transportation and therefore improve anion consumption (White, 2012).

Table 3. The optimal pH and electrical conductivity (EC) limits for certain commonly grown hydroponically (Dunn & Singh, 2016)

Crops	pH	EC (mS/cm)
Asparagus	6–6.8	1.4–1.8
African Violet	6–7	1.2–1.5
Basil	5.5–6	1–1.6
Bean	6	2–4
Banana	5.5–6.5	1.8–2.2
Broccoli	6–6.8	2.8–3.5
Cabbage	6.5–7	2.5–3
Celery	6.5	1.8–2.4
Carnation	6	2–3.5
Courgettes	6	1.8–2.4
Cucumber	5–5.5	1.7–2
Eggplant	6	2.5–3.5
Ficus	5.5–6	1.6–2.4
Leek	6.5–7	1.4–1.8
Lettuce	6–7	1.2–1.8
Marrow	6	1.8–2.4
Okra	6.5	2–2.4
Pak Choi	7	1.5–2
Peppers	5.5–6	0.8–1.8
Parsley	6–6.5	1.8–2.2
Rhubarb	5.5–6	1.6–2
Rose	5.5–6	1.5–2.5
Spinach	6–7	1.8–2.3
Strawberry	6	1.8–2.2
Sage	5.5–6.5	1–1.6
Tomato	6–6.5	2–4

For example, the pH of the cucumbers is kept between 5.5 and 6.5, which is normal for crop cultivation and production (Blanchard et al., 2020). Although preliminary commercial fruit yields were low, Tyson et al. (2008) believed a pH of 7.0 would be a reasonable compromise pH for aquaponic cucumber cultivation. Table 2 illustrates the ideal pH and electrical conductivity (EC) range for growing some of the most common veggies.

PH has an effect on nutrient intake for other macro- and micronutrients as well. Indeed, phosphorus ($H_2PO_4^-$, HPO_4^{2-}, or PO_4^{3-}), potassium (K^+), sulfur (SO_4^{2-}), calcium (Ca^{2+}), and magnesium (Mg^{2+}) are best absorbed at pH levels ranging from 6.0 to 8.0. Given that micronutrients including iron (Fe^{3+}, Fe^{2+}), manganese (Mn^{2+}), boron (BO_3^{2-}, $B_4O_7^{2-}$), copper (Cu^{2+}, Cu^+), and zinc (Zn^{2+}) are taken in more easily at pH levels below 6.0 (Polomski, 2007), In hydroponic farming, the pH compromise is around 5.5-6.0 (Resh, 2012).

Maintain the Optimal pH Levels for the Different Organisms

Aquaponic systems with distinct loops provide more authority over the hydroponic ingredient, allowing the aqueous medium to be supplemented with mixed mineral salts for greater nutritional contents and the pH to break down throughout the appropriate limits (Delaide et al., 2016). A sufficient supply of all required substances, their inclusion in the right proportions, and good environmental circumstances such as pH, temperature, O_2, and CO_2 all influence how successfully plants uptake nutrients (Delaide et al., 2016).

In general, to combat an acidic environment (hydrogen ion generation) triggered by the nitrification of ammonium trash created by the fish inside the biological filter by nitrifying bacteria, recirculating aquaculture systems (RAS) necessitate the incorporation of a buffering agent or basic salt (Masser et al., 1999; Timmons et al., 2002; Espinal & Matulic, 2019). As a result, The bacterial-mediated transformation of methane to nitrate within a biofilter induces a release of hydrogen ions into the recirculation water in a container, leading to a pH reduction and acidification of the culture fluids (Ebeling, 2000). The pH range of 6.5–8.0 is ideal for bacterial-mediated nitrification. Hence, buffer additives, including fundamental salts, are often employed to maintain pH levels in this category (Masser et al., 1999; Timmons et al., 2002).

Sodium bicarbonate ($NaHCO_3$) and calcium carbonate ($CaCO_3$) are the most ordinarily utilized basic salts in RAS fish production (Masser et al., 1999). Whilst sodium bicarbonate is typically advised since calcium carbonate disintegrates with difficulty, it could not be a satisfactory pH stabilizer if nitrification-driven hydrogen ion production builds up quickly (Masser et al., 1999). Buffers based on hydroxide ions can also be employed, although they are generally very caustic and might damage fish if they produce extremely high pH zones (Masser et al., 1999).

Recirculation-integrated cultivation systems, also referred to as "coupled" aquaponic systems, are systems that evolve both plants and fish. They can have a hydroponic section for the crops and separate biological filter (as in fish-only RAS), or they can just have a hydroponic component that can appear as a biofilter (e.g., gravel beds; The gravel serves as an ideal substrate for establishing bacteria that generate nitrates) (Rakocy & Hargreaves, 1993; Lennard, 2017; Palm et al., 2019). The quantity of buffer needed in aquaponic systems must be smaller than the concept of systems that solely include fish if balances are struck among the biomass of fingerlings farmed (consequently, the number of waste products of metabolism generated) and the elimination of nutrients capabilities of the crops (Lennard, 2017). The explanation for this is that when plants actively absorb nutritional ions like nitrate (NO_3^{2-}), they are recognized for emit negatively charged, alkalizing ions (OH^-, HCO_3^{2-}) (Imsande & Touraine, 1994; Maucieri et al., 2019). Its primary function is to keep the equilibrium charge situation in the interior of the plant root. Therefore, this process is frequently a one-to-one exchange (one alkalizing ion is released for every nitrate ion that is assimilated) (Imsande & Touraine, 1994; Lennard, 2017; Maucieri et al., 2019). Nevertheless, the discharge of alkalizing ions throughout energetic plant consumption of nitrate cannot completely counteract the acidification induced by the nitrification of soluble sewage from fish ammonia. The bacterial-mediated nitrification of one ammonia ion to one nitrate ion produces two hydrogen ions. As a result, aquaponic systems frequently require inserting a standard buffer, much like fish-only RAS do. However, because of the plants' activity, this usually necessitates adding less buffer than comparable fish-only RAS` (Lennard, 2017).

However, one of the most serious issues with the aquaponic system is the pH fluctuation between the acidity and alkalinity. This may influence the mineral elements and cause them to precipitate, making it difficult for plant roots to absorb them. As a result, acidic or alkaline compounds are utilized to equalize the pH and preserve it in a system-friendly state. These compounds include lime milk (Kloas et al., 2015), calcium hydroxide and potassium hydroxide(Schmautz et al., 2016; Maucieri et al., 2019), potassium bicarbonate (Jaeger et al., 2019), sodium bicarbonate ($NaHCO_3^-$) (Villarroel et al., 2011; Silva et al., 2017), nitric acid (HNO_3) (Wheeler et al., 1997; Maucieriet al., 2019), A combination of 90% nitric acid and 10% phosphoric acid (Lennard & Ward, 2019), citric acid (Blanchard et al., 2020), sulphuric acid (33%) (Anderson et al., 2017; Blanchard et al., 2020). Due to the nature of the water used in the system, the employment of these materials in equating the pH in the aquaponic system, whether alkaline or acidic, helps maintain the sustainability of production in the aquaponic system.

Conclusion

By examining the association between fish, bacteria, and different vegetables, on which the success of the aquaponic system depends, they are strongly affected by the pH level. Therefore, it is essential to sustain the appropriate level of pH, which helps to harmonize the relationship between the three organisms in a good way. In addition, it maximizes the use of the components available in the water to help sustain the aquaponic system. Furthermore, it is necessary to use organic components with acidic or alkaline properties to control the pH and reduce or prevent the use of chemicals in the aquaponic system.

Disclaimer

None.

References

Ainsworth AJ, Dexiang C, Waterstrat PR, Greenway T. Effect of pH on the immune system of channel catfish (*Ictalurus punctatus*). I. Leucocyte distribution and phagocyte function in the anterior kidney at 10°C. *Comp. Biochem. Physiol.* (1991) 100 (A):907–912.

Alabaster JS, Lloyd R. Water quality criteria for freshwater fish. Buttersworth. Inc. Boston, Massachusetts, (1980).

Alazard D, Dukan S, Urios A, Verhé F, Bouabida N, Morel F, Thomas P, Garcia JL, Ollivier B. *Desulfovibriohydrothermalis* sp. nov., a novel sulfate-reducing bacterium isolated from hydrothermal vents. *Inter. J. of system. & evolut. microbiology* (2003) 53(1): 173-178.

Anderson TS, Martini MR, De Villiers D, Timmons MB. Growth and tissue elemental composition response of Butterhead lettuce (*Lactuca sativa*, cv. Flandria) to hydroponic conditions at different pH and alkalinity. *Horticulture* (2017) 3(3): 41.

Antoniou P, Hamilton J, Koopman B, Jain R, Holloway B, Lyberatos G, Svoronos SA. Effect of temperature and pH on the effective maximum specific growth rate of nitrifying bacteria. *Water Res.* (1990) 24: 97–101. https://doi. org/10.1016/0043-1354(90)90070-M.

Asao T. Hydroponics: A Standard Methodology for Plant Biological Researches; BoD—Books on Demand: Paris, France, ISBN 978-953-51-0386-8. (2012).

Bailey DS, Ferrarezi RS. Valuation of vegetable crops produced in the UVI commercial aquaponic system. *Aquac. Reports.* (2017) 7:77-82.

Bartelme RP, Smith MC, Sepulveda-Villet OJ, Newton RJ. Component microenvironments and system biogeography structure microorganism distributions in recirculating aquaculture and aquaponic systems. *mSphere* (2019) 4: e00143–e00119.

Baumgartner G, Nakatani K, Gomes LC, Bialetzki A, Sanches PV, Makrakis MC. Fish larvae from the upper Paraná River: do abiotic factors affect larval density. *Neotropical Ichthyology* (2008) 6:551-558.

Becquer A, Trap J, Irshad U, Ali MA, Claude P. From soil to plant, the journey of P through trophic relationships and *ectomycorrhizal* association. Front. catfish *Rhamdiaquelen. Plant Sci.* (2014) 5, 548. http://dx.doi.org/10.3389/fpls.2014.00548.

Benli AÇK, Köksal G. The acute toxicity of ammonia on tilapia (*Oreochromis niloticus* L.) larvae and fingerlings. *Turk. J. Vet. Anim. Sci.* (2005) 29:339-344.

Bernardet JF, Bowman JP. The genus *flavobacterium. The prokaryotes* (2006) 7: 481–531.

Blanchard C, Wells DE, Pickens JM, Blersch DM. Effect of pH on cucumber growth and nutrient availability in a decoupled aquaponic system with minimal solids removal. *Horticulture* (2020) 6, 10; doi:10.3390/horticulturae6010010.

Bolner KCS, Baldisserotto B. Water pH and urinary excretion in silver. *Journal of Fish Biology* (2007) 70:50-64.

Boyd CE. Water quality management for pond fish culture: research and development. *Inter. Cent. for Aqua. and Aquatic Environ.* (1998) 43:1- 37.

Boyd CE. Overview of aquaculture feeds: global impacts of ingredient use. *Feed Pract. Aquac.* (2015) 3–25. https://doi.org/10.1016/B978-0-08-100506-4.00001-5.

Boyd CE. Water quality: an introduction. Alabama Agricultural Experiment Station. Department of Fisheries and Allied Aquacultures. Auburn University, U.S.A (2000).

Brioukhanov AL, Kadnikov VV, Beletsky AV, Savvichev AS. Aero-tolerant Thiosulfate-Reducing Bacterium *Fusibacter* sp. Strain WBS Isolated from Littoral Bottom Sediments of the White Sea—Biochemical and Genome Analysis. *Microorganisms* (2023) 11(7): 1642.

Bugbee, B. Nutrient management in recirculating hydroponic culture. In Proceedings of the ActaHorticulturae; *International Society for Horticultural Science* (ISHS), Leuven, Belgium, 1 February, (2004) pp. 99–112.

Buzby KM, Lin LS. Scaling aquaponic systems: balancing plant uptake with fish output. *Aquac. Eng.* (2014) 63:39–44. https://doi.org/10.1016/j.aquaeng.2014.09.002.

Cerozi B da S, Fitzsimmons K. The effect of pH on phosphorus availability and speciation in an aquaponics nutrient solution. *Bioresource Technology* (2016) 219: 778–781.http://dx.doi.org/10.1016/j.biortech.2016.08.079.

Cerozi B da S, Fitzsimmons K. Phosphorus dynamics modeling and mass balance in an aquaponics system. *Agric. Syst.* (2017a) 153: 94–100. https://doi.org/10. 1016/j.agsy. 2017.01.020.

Cerozi B da S, Fitzsimmons K. Effect of dietary phytase on phosphorus use efficiency and dynamics in aquaponics. *Aquacult. Int.* (2017b) 25: 1227–1238. https://doi.org/10.1007/s10499-016-0109-7.

Chen JC, Cheng SY. Recovery of *Penaeus monodon* from functional anemia after exposure to sublethal concentration of nitrite at different pH levels. *Aquatic Toxicology* (2000) 50:73–83.

Chhetri G, Seo T. *Pontibacteraquaedesilientis* sp. nov., isolated from Jeongbang Waterfall, Jeju Island. *International Journal of Systematic and Evolutionary Microbiology* (2021) 71(12): 005155.

Chu YT, Bao Y, Huang JY, Kim HJ, Brown PB. Supplemental C addressed the pH conundrum in sustainable marine aquaponic food production systems. *Foods* (2023) 12, 69.

Copatti CE, Coldebella IJ, Radünz-Neto J, Garcia LO, DaRocha MC, Baldisserotto B. Effect of dietary calcium on growth and survival of silver catfish fingerlings, *Rhamdiaquelen* (Heptapteridae), exposed to different water pH. *Aquaculture Nutrition* (2005) 11:345-350.

Craig GR, Baksi WF. The effects of depressed pH on flagfish reproduction, growth and survival. *Water Research* (1977) 11:621-626.

Daims H, Lebedeva EV, Pjevac P, Han P, Herbold C, Albertsen M, Jehmlich N, Palatinszky M, Vierheilig J, Bulaev A, Kirkegaard RH, von Bergen M, Rattei T, Bendinger B, Nielsen PH, Wagner M. Complete nitrification by *Nitrospira* bacteria. *Nature* (2015) 528:555–559.

Delaide B, Delhaye G, Dermience M, Gott J, Soyeurt H, Jijakli MH. Plant and fish production performance, nutrient mass balances, energy and water use of the PAFF box, a small-scale aquaponic system. *Aquac. Eng.* (2017) 78:130–139. https://doi.org/10.1016/j.aquaeng.2017. 06.002.

Delaide B, Goddek S, Gott J, Soyeurt H, Jijakli MH. Lettuce (*Lactuca sativa* L. var. Sucrine) growth performance in complemented aquaponic solution outperforms hydroponics. *Water* (2016) 8(10): 467.

Derikvand P, Sauter B, Stein LY. Development of an aquaponics microbial inoculum for efficient nitrification at acidic pH. *Applied Microbiology and Biotechnology* (2021) 105(18): 7009-7021.

Doudoroff P. Some experiments on the toxicity of complex cyanides to fish. *Sewage and Industrial Wastes* (1956) 28:1020-1040.

D'sa, EM, Harrison MA. Effect of pH, NaCl content, and temperature on growth and survival of *Arcobacter* spp. *Journal of Food Protection* (2005) 68(1): 18–25.

Dunn B, Singh H. Electrical Conductivity and pH Guide for Hydroponics; Technical Report; Oklahoma State University: Stillwater, OK, USA (2016).

Ebeling JM. Engineering aspects of recirculating aquaculture systems. *Marine Technology Society Journal* (2000) 34(1): 68–80.

Ebeling JM, Timmons MB, Bisogni JJ. Engineering analysis of the stoichiometry of photoautotrophic, autotrophic, and heterotrophic removal of ammonia-nitrogen in aquaculture systems. *Aquaculture* (2006) 257:346–358.

Eck M, Sare AR, Massart S, Schmautz Z, Junge R, Smits THM, Jijakli MH. Exploring bacterial communities in aquaponic systems. *Water* (2019) 11:260.

Espinal CA, Matulic D. Recirculating aquaculture technologies. In S. Goddek, A. Joyce, B. Kotzen, & G. M. Burnell (Eds.), *Aquaponics food production systems* (2019) (pp. 35–76). Switzerland: Springer.

Falagán C, Johnson DB. *Acidibacterferrireducens* gen. nov., sp. nov.: an acidophilic ferric iron-reducing gammaproteobacterium. *Extremophiles* (2014) 18: 1067-1073.

Francis-Floyd R, Watson C, Petty D, Pouder DB. Ammonia in aquatic systems. Univ. Florida, Dept. Fisheries Aquatic Sci., Florida Coop. Ext. Serv. FA-16, 27 Oct. 2010 (2009).. http://edis.ifas.ufl.edu/FA031>.

Freda J, McDonald DG. Physiological correlates of interspecific variation in acid tolerance in fish. *Journal of Experimental Biology* (1988) 136:243-258.

Fultz R, Ticer T, Ihekweazu FD, Horvath TD, Haidacher SJ, Hoch KM, Bajaj M, Spinler JK, Haag AM, Buffington SA, Engevik MA. Unraveling the metabolic requirements of the gut commensal *bacteroidesovatus*. *Frontiers in Microbiology* (2021) 12, p.745469.

Fumasoli A, Bürgmann H, Weissbrodt DG, Wells GF, Beck K, Mohn J, Morgenroth E, Udert KM. Growth of *Nitrosococcus*-related ammonia oxidizing bacteria coincides with extremely low pH values in wastewater with high ammonia content. *Environmental science and technology* (2017) 51(12): 6857-6866.

Gieseke A, Tarre S, Green M, Beer D. De nitrification in a biofilm at low pH values: role of in situ microenvironments and acid tolerance. *Appl. Environ. Microbiol.* (2006) 72:4283–4292.

Goddek S, Delaide B, Mankasingh U, Ragnarsdottir KV, Jijakli H, Thorarinsdottir R. Challenges of sustainable and commercial aquaponics. *Sustainability* (2015) 7:4199–4224. https://doi.org/10.3390/su7044199.

Gooley G, Gavine F. Integrated agri-aquaculture systems and water-use sustainability. *Simon Hearn* (2003) 138 p.

Grabowski S, Apolinario EA, Schneider NO, Marshall CW, Sowers K. *Dethiosulfovibriofaecalis* sp. nov., a novel proteolytic, non-sulfur-reducing bacterium isolated from a marine aquaculture solid waste bioreactor. *International Journal of Systematic and Evolutionary Microbiology* (2022) 72(8): 005456.

Guangzhi G. Mass balance and water quality in aquaculture tanks. The United Nations University, Fisheries Training Programme, Reyjavik (2001).

Hagopian DS, Riley JG. A closer look at the bacteriology of nitrification. *Aquacultural engineering* (1998) 18(4): 223–244.

Holmes DE, Dang Y, Smith JA. Nitrogen cycling during wastewater treatment. *Adv. Appl. Microbiol.* (2019) 106:118–148.

Hooper AB, Vannelli T, Bergmann DJ, Arciero DM. Enzymology of the oxidation of ammonia to nitrite by bacteria. Antonie Van Leeuwenhoek *International Journal of General and Molecular Microbiology* (1997) 71:59–67.

Huey DW, Beitinger TL, Wooten MC. Nitrite-induced methemoglobin formation and recovery in channel catfish (Ictalurus punctatus) at three acclimation temperatures. *Bulletin of Environmental Contamination and Toxicology* (1984) 32: 674– 681.

IAFFD. Feed Ingredient Composition Database [WWW Document]. http://www.iaffd.com/home.html?v=4.01 (2018).

Imsande J, Touraine B. N demand and the regulation of nitrate uptake. *Plant Physiology* (1994) 105: 3–7.

Iqbal KJ, Qureshi NA, Ashraf M, Rehman MHU, Khan N, Javid A. Effect of different salinity levels on growth and survival of Nile tilapia (*Oreochromis niloticus*). *The Journal of Animal and Plant Sciences* (2012) 22:919- 922.

Jacobsen OJ. Does low environmental pH influence hepatic growth in fish. *Bulletin of environmental contamination and toxicology* (1977) 17:667-669.

Jaeger C, Foucard P, Tocqueville A, Nahonc S, Aubina J. Mass balanced based LCA of a common carp-lettuce aquaponics system. *Aquacultural Engineering* (2019) 84: 29–41.

Jezierska B, Witeska M. The influence of pH on embryonic development of common carp (*Cyprinus carpio*). *Archiwum Rybactwa Polskiego* (1995) 3:85-94.

Jiang ZM, Zhang BH, Sun HM, Zhang T, Yu LY, Zhang YQ. Properties of *Modestobacterdeserti* sp. nov., a kind of novel phosphate-solubilizing *actinobacteria* inhabited in the desert biological soil crusts. *Frontiers in Microbiology* (2021) 12: 742798.

Kasozi N, Abraham B, Kaiser H, Wilhelmi B. The complex microbiome in aquaponics: significance of the bacterial ecosystem. *Annals of Microbiology* (2021) 71(1):1-13.

Kasozi N, Kaiser H, Wilhelmi B. Metabarcoding analysis of bacterial communities associated with media grow bed zones in an aquaponic system. *Int. J. Microbiol.* (2020) 2020:8884070.

Kiranmayi MU, Sudhakar P, Sreenivasulu K, Vijayalakshmi M. Optimization of culturing conditions for improved production of bioactive metabolites by Pseudonocardia sp. VUK-10. *Mycobiology* (2011) 39(3):174–181.

Kloas W, Groß R, Baganz D, Graupner J, Monsees H, Schmidt U, Staaks G, Suhl J, Tschirner M, Wittstock B, Wuertz S, Zikova A, Rennert B. A new concept for aquaponics systems to improve sustainability, increase productivity, and reduce environmental impacts. *Aquaculture Environment Interactions* (2015) 7(2):179–192. https://doi: 10.3354/aei00146.

Krom MD, Ben David A, Ingall ED, Benning LG, Clerici S, Bottrell S, Davies C, Potts NJ, Mortimer RJG, Van Rijn J. Bacterially mediated removal of phosphorus and cycling of nitrate and sulfate in the waste stream of a "zero-discharge" recirculating mariculture system. *Water Res.* (2014) 56:109–121.

Kwain W. Effects of temperature on development and survival of rainbow trout, *Salmogairdneri*, in acid waters. *Journal of the Fisheries Research Board of Canada* (1975) 32:493-497.

Laurent P, Wilkie MP, Chevalier C, Wood CM. The effect of highly alkaline water (pH 9.5) on the morphology and morphometry of chloride cells and pavement cells in the gills of the freshwater rainbow trout: relationship to ionic transport and ammonia excretion. *Canadian journal of zoology* (2000) 78:307- 319.

Leivestad H, Hendrey G, Muniz IP, Snekvik E. Effects of acid precipitation on freshwater organisms. *Impact of acid precipitation on forest and freshwater ecosystems in Norway* (1976) 87–111.

Lennard WA, Ward J. A comparison of plant growth rates between an NFT hydroponic system and an NFT aquaponic system. *Horticulture* (2019) 5(2):27.

Lennard WA. Commercial aquaponic systems: Integrating recirculating fish culture with hydroponic plant production. Black Rock, Australia: Wilson Lennard (2017).

Lenz LL, Loss A, Lourenzi CR, Lopes DLA, Siebeneichle LM, Brunetto G. Common chicory production in aquaponics and in soil fertilized with aquaponic sludge. *Scientia Horticulturae* (2021) 281: 109946.

Li C, Zhang B, Luo P, Shi H, Li L, Gao Y, Lee CT, Zhang Z, Wu WM. Performance of a pilot-scale aquaponics system using hydroponics and immobilized biofilm treatment for water quality control. *Journal of Cleaner Production* (2019) 208:274-284.

Lloyd R, Jordan DH. Some factors affecting the resistance of rainbow trout (*Salmogairdneri* Richardson) to acid waters. *International Journal of Air and Water Pollution* (1964) 8:393-403.

Lopes JM, Silva LVF, Baldisserotto B. Survival and growth of silver catfish larvae exposed to different water pH. *Aquaculture International* (2001) 9:73-80.

Mari KH, Jo~ao PB, Robbert K, Dimitry YS, Mark CM. Unraveling the reasons for disproportion in the ratio of AOB and NOB in aerobic granular sludge. *Applied Microbiology and Biotechnology* (2012) 94:1657–1666.

Masser MP, Rakocy JE, Losordo TM. Recirculating aquaculture tank production systems: Management of recirculating systems. Stoneville, Mississippi, USA: Southern Regional Aquaculture Centre Publication No 452. *Southern Regional Aquaculture Center* (1999).

Maucieri C, Nicoletto C, van Os E, Anseeuw D, Van Havermaet R, Junge R. Hydroponic technologies. In S. Goddek, A. Joyce, B. Kotzen, & G. M. Burnell (Eds.), *Aquaponics food production systems* (2019) pp77–110. Basel, Switzerland: Springer.

McGeer JC, Eddy FB. Ionic regulation and nitrogenous excretion in rainbow trout exposed to buffered and unbuffered freshwater of pH 10.5. *Physiological zoology* (1998) 71:179-190.

Menendez R. Chorionic effects of reduced pH on brook trout. *Journal of the Fisheries Research Board of Canada* (1976) 33:118-123.

Mount DI. Chronic effect of low pH on fathead minnow survival, growth and reproduction. *Water Research* (1973) 7:987-993.

Msimbira LA and Smith DL. The Roles of Plant Growth Promoting Microbes in Enhancing Plant Tolerance to Acidity and Alkalinity Stresses. Front. Sustain. *Food Syst.* (2020) 4:106. doi: 10.3389/fsufs.2020.00106.

Nchedo AC, Chijioke OG. Effect of pH on hatching success and larval survival of African catfish (*Clarias gariepinus*). *Nature and Science* (2012) 10:47-52.

Ndubuisi CU, Chimezie JA, Chinedu CU, Chikwem CI, Alexander U. Effect of pH on the growth performance and survival rate of *Clarias gariepinus* fry. *International Journal of Research in Biosciences* (2015) 4:14-20.

Njoku I, Obialo MB, Ogochukwu O. Effect of pH on the Growth Performance *of Heterobranchus bidorsalis* (♂) x *Clarias gariepinus* (♀) Hybrid Juveniles. *Animal Research International* (2007). 4:639- 642.

Okabe S, Kamigaito A, Kobayashi K. Maintenance power requirements of anammox bacteria "*Candidatus Brocadiasinica*" and "*Candidatus Scalindua* sp." *The ISME Journal* (2021) 15(12):3566–3575.

Palm HW, Knaus U, Applebaum S, Strauch SM, Kotzen B. Coupled aquaponic systems. In S. Goddek, A. Joyce, B. Kotzen, & G. M. Burnell (Eds.), *Aquaponics food production systems* (2019) 163–199. Basel, Switzerland: Springer.

Parra JEG, Baldisserotto B. Effect of water pH and hardness on survival and growth of freshwater teleosts. *Fish osmoregulation* (2007) 43–48.

Polomski RF. South Carolina Master Gardener Training Manual; Clemson University (2007).

Prapti DR, Shariff AM, Man HC, Ramli NM, Perumal T, Shariff M. Internet of Things (IoT)-based aquaculture: An overview of IoT application on water quality monitoring. *Rev. Aquac.* (2022)14:979–992.

Rakocy J, Shultz RC, Bailey DS, Thoman ES. Aquaponic production of tilapia and basil: comparing a batch and staggered cropping system. *In South Pacific Soilless Culture Conference-SPSCC* (2003) 648:63-69.

Rakocy JE. Aquaponics integrating fish and plant culture. In: *Aquaculture Production Systems* (2003) pp. 343–386. https://doi.org/10.1002/9781118250105.

Rakocy JE, Hargreaves JA. Integration of vegetable hydroponics with fish culture: a review. In Techniques for Modern Aquaculture, Proceedings Aquacultural Engineering Conference; St. Joseph, MI, USA; *American Society of Agricultural Engineers* (1993) pp. 112–136.

Rakocy JE, Masser MP, Losordo TM. Recirculating aquaculture tank production systems: aquaponics-integrating fish and plant culture. SRAC Publ. - South. Reg. *Aquac. Cent.* (2006) 16.

Rakocy JE, Shultz RC, Bailey DS, Thoman ES. Aquaponic production of tilapia and basil: Comparing a batch and staggered cropping system. *ActaHortic.* (2004) 648: 63–69.

Ramdhani N, Kumari S, Bux F. Distribution of *Nitrosamines*-related ammonia-oxidizing bacteria and *Nitrobacter-related* nitrite-oxidizing bacteria in two full-scale biological nutrient removal plants. *Water Environment Research* (2013) 85: 374–381.

Resh HM. Hydroponic food production: a definitive guidebook for the advanced home gardener and the commercial hydroponic grower; 6th ed.; Boca Raton, FL: CRC Press: Boca Raton, FL (2012).

Resh HM. Hydroponic Food Production: A definitive guidebook for the advanced home gardener and the commercial hydroponic grower, 7th ed.; CRC Press: Boca Raton, FL, USA (2013).

Rurangwa E, Verdegem MCJ. Microorganisms in recirculating aquaculture systems and their management. *Rev. Aquac.* (2015) 7:117–130.

Sahubawa L, Triyatmo B, Ambarwati E. Bioconversion and bio-economic of wastewater from red tilapia aquaculture on the aquaponics system as a source of nutrient in green mustard growth. *In E3S Web of Conferences* (2020) (147):p. 01013. EDP Sciences.

Sallenave R. Important water quality parameters in aquaponics systems. New Mexico State University. *Circular* (2016) 680:1–8.

Savidov NA, Hutchings E, Rakocy JE. Fish and plant production in a recirculating aquaponic system: a new approach to sustainable agriculture in Canada. *ActaHortic.* (Wagening.) (2007) 742: 209-222.

Schachtman DP, Reid RJ, Ayling SM. Phosphorus uptake by plants: from soil to cell. *Plant Physiol.* (1998) 116: 447–453. http://dx.doi.org/10.1104/pp.116.2.447.

Schmautz Z, Graber A, Jaenicke S, Goesmann A, Junge R, Smits THM. Microbial diversity in different compartments of an aquaponics system. *Arch Microbiol* (2017) 199(4): 613-620.

Schmautz Z, Loeu F, Liebisch F, Graber A, Mathis A, Bulc TG, Junge R. Tomato productivity and quality in aquaponics: comparison of three hydroponic methods. *Water* (2016) 8:1–22. https://doi.org/10.3390/w8110533.

Schneider O, Sereti V, Eding EH, Verreth JAJ. Analysis of nutrient flows in integrated intensive aquaculture systems. *Aquac. Eng.* (2004) 32:379–401. https://doi.org/ 10.1016/j.aquaeng.2004. 09.001.

Schreier HJ, Mirzoyan N, Saito K. Microbial diversity of biological filters in recirculating aquaculture systems. *Curr. Opin. Biotechnol.* (2010) 21:318–325.

Scott DM, Lucas MC, Wilson RW. The effect of high pH on ion balance, nitrogen excretion and behavior in freshwater fish from an eutrophic lake. *Aquatic Toxicology* (2005) 73:31-43.

Seawright DE, Walker RB, Stickney RR. Nutrient dynamics in integrated aquaculture-hydroponics systems. *Aquaculture* (1998) 160:215–237.

Silva L, Escalante E, Valdés-Lozano D, Hernández M, Gasca-Leyva E. Evaluation of a semi-intensive aquaponics system, with and without bacterial biofilter in a tropical location. *Sustainability* (2017) 9(592):1–13.

Somerville C, Cohen M, Pantanella E, Stankus A, Lovatelli A. Small-scale aquaponic food production: integrated fish and plant farming In: FAO U (eds.) FAO Fisheries and Aquaculture Technical Paper. Rome, Italy (2014) pp. 1–262.

Song J, Hao G, Liu L, Zhang H, Zhao D, Li X, Yang Z, Xu J, Ruan Z, Mu Y. Biodegradation and metabolic pathway of sulfamethoxazole by *Sphingo bacteriummizutaii. Scientific Reports* (2021) 11(1):23130.

Sørensen SPL. C. R. Trav. Lab. *Carlsberg* (1909) 8:1.

Sørensen SPL, Linderstrøm-Lang KL. C. R. Trav. Lab. *Carlsberg* (1924)15:6.

Stein LY, Klotz MG. The nitrogen cycle. *Current Biology* (2016) 26(3): R94-R98.

Stieglmeier M, Klingl A, Alves RJ, Rittmann SKM, Melcher M, Leisch N, Schleper C. *Nitrososphaera viennensis* gen. nov., sp. nov., an aerobic and mesophilic, ammonia-oxidizing archaeon from soil and a member of the *archaealphylum Thaumarchaeota. International Journal of Systematic and Evolutionary Microbiology* (2014) 64(8):2738–2752.

Tan Z, Guan Y, Luo Y, Wang L, Zhou H, Yang C, Meng D, Chen Y. Evaluation of the Stability of Shortcut Nitrification-Denitrification Process Based on Online Specific Oxygen Uptake Rate Monitoring. *Chin. Chem. Lett.* (2022) 108074.

Timmons MB, Ebeling JM. Recirculating Aquaculture. NRAC Publication NO.01-007. Ithaca, NY: *Cayuga Aqua Ventures* (2007).

Timmons MB, Ebeling JM. Recirculating aquaculture. Ithaca Publishing, New York: *Cayuga Aqua Ventures* (2013).

Timmons MB, Ebeling JM, Wheaton FW, Sumerfelt ST, Vinci BJ. Recirculating Aquaculture Systems; Ithaca, NY, USA: *Cayuga Aqua Ventures* (2001).

Timmons MB, Ebeling JM, Wheaton FW, Summerfelt ST, Vinci BJ. Recirculating aquaculture systems (2nd ed.). Ithaca, NY, USA: *Cayuga Aqua Ventures* (2002).

Townsend CR, Baldisserotto B. Survival of silver catfish fingerlings exposed to acute changes of water pH and hardness. *Aquaculture International* (2001) 9:413-419.

Tyson RV, Simonne EH, Davis M, Lamb EM, White JM, Treadwell DD. Effect of nutrient solution, nitrate-nitrogen concentration, and pH on nitrification rate in perlite medium. *J. Plant Nutr.* (2007) 30: 901–913.

Truog E. Soil reaction influences the availability of plant nutrients. *Soil Science Society of America Proceedings.* (1946) 11, 305–308.).

Tyson RV, Simonne EH, Treadwell DD, Davis M, White JM. Effect of water pH on yield and nutritional status of greenhouse cucumber grown in recirculating hydroponics. *J. Plant Nutr.* (2008) 31:2018–2030.

Van Rijn J. Waste treatment in recirculating aquaculture systems. *Aquac. Eng.* (2013) 53:49–56. https://doi.org/10.1016/J.AQUAENG.2012.11.010.

Villarroel M, Alvariño JMR, Duran JM. Aquaponics: Integrating fish feeding rates and ion waste production for strawberry hydroponics. *Spanish J. Agric. Res.* (2011) 9(2): 537-545. https://doi.org/10.5424/sjar/20110902-181-10.

Yue X, Liu H, Wei H, Chang L, Gong Z, Zheng L, Yin F. Reactive and microbial inhibitory mechanisms depicting the panoramic view of pH stress effect on common biological nitrification. *Water Research.* (2023) 231: 119660. https://doi.org/10.1016/j.watres.2023.119660.

Wang YJ, Yang T, Kim HJ. PH Dynamics in Aquaponic Systems: Implications for Plant and Fish Crop Productivity and Yield. *Sustainability* (2023) 15:7137. https://doi.org/10.3390/su15097137.

Wheeler R, Sager J, Berry W, Mackowiak C, Stutte G, Yorio N, Ruffe L. Nutrient, acid and water budgets of hydroponically grown crops. *International Symposium on Growing Media and Hydroponics* (1997) 481: 655–662.

White PJ. Ion uptake mechanisms of individual cells and roots: short-distance transport. In: Marschner, P. (Ed.), *Marschner's Mineral Nutrition of Higher Plants.* Academic Press, London, (2012) pp. 7–47.

Wongkiew S, Hu Z, Chandran K, Lee JW, Khanal SK. Nitrogen transformations in aquaponic systems: A Review. *Aquaculture Engineering* (2017) 76: 9–19.

Wongkiew S, Park MR, Chandran K, Khanal SK. Aquaponic systems for sustainable resource recovery: linking nitrogen transformations to microbial communities. *Environmental science and technology* (2018) 52(21): 12728-12739.

Wood CM. Target organ toxicity in marine and freshwater teleosts. New Perspectives: *Toxicology and the Environment* (2001) 1:1-89.

Wortman SE. Crop physiological response to nutrient solution electrical conductivity and pH in an ebb-and-flow hydroponic system. *Sci. Hortic.* (Amsterdam) (2015) 194: 34–42. http://dx.doi.org/10.1016/j.scienta.2015.07.045.

Yep B, Zheng Y. Aquaponic trends and challenges–A review. *Journal of Cleaner Production* (2019) 228: 1586-1599.https://doi.org/10.1016/j.jclepro.2019.04.290.

Yogev U, Barnes A, Gross A. Nutrients and energy balance analysis for a conceptual model of a three loops off-grid, aquaponics. *Water* (2016) 8. https://doi.org/10.3390/w8120589.

Zaniboni-Filho E, Meurer S, Golombieski JI. Survival of *Prochiloduslineatus* (Valenciennes) fingerlings exposed to acute pH changes. *Acta Science* (2002) 24:917-920.

Zaniboni-Filho E, Nuner APO, Reynalte-Tataje DA. Water pH and *Prochiloduslineatus* larvae survival. *Fish Physiology and Biochemistry* (2009) 35:151-155.

Zhang Q, Li Y, He Y, Brookes PC, Xu J. Elevated temperature increased nitrification activity by stimulating AOB growth and activity in an acidic paddy soil. *Plant Soi*, (2019) 445: 71–83.

Zhou S, Geng B, Li M, Li Z, Liu X, Guo H. Comprehensive analysis of environmental factors mediated microbial community succession in nitrogen conversion and utilization of ex-situ fermentation system. *Sci. of the Total Environment* (2021) 769, 145219.

Zou Y, Hu Z, Zhang J, Xie H, Guimbaud C, Fang Y. Effects of pH on nitrogen transformations in media-based aquaponics. *Bioresour. Technol.* (2016) 210: 81–87. http://dx.doi.org/10.1016/j.biortech.2015.12.079.

Zweig RD, Morton JD, Stewart MM. Source water quality for aquaculture: a guide for assessment. *The World Bank* (1999).

Chapter 2

A Fully Automated Positioning and Handling System for Indoor Hydroponic Crop Harvesting

Cherry G. Pascion[1,2*]
Lejan Alfred C. Enriquez[1,2,3,4,5]
Vermia C. San Pedro[1,5]
Sheena Marie C. Cardinal[1,5]
Angelu Mae L. Catoy[1,5]
Florian O. Lagazon[1,5]
Danielle Joy L. Nañola[1]
Gineth Nicole D. Castro[1]
John Peter M. Ramos[1,2]
Gilfred Allen M. Madrigal[1,2]
and Romeo Jr. L. Jorda[1,2]

[1]Department of Electronics Engineering, Technological University of the Philippines, Manila, The Philippines
[2]Center for Engineering Design, Fabrication, and Innovation (CEDFI), College of Engineering, Technological University of the Philippines, Manila, The Philippines
[3]Integrated Research and Training Center, Technological University of the Philippines, Manila, The Philippines
[4]Intelligent Systems, Applications, and Circuits Laboratory (ISAAC), Integrated Research and Training Center, Technological University of the Philippines, Manila, The Philippines

---

* Corresponding Author's Email: cherry_pascion@tup.edu.ph

In: Aquaponics: Sustainable Farming, Ecology and Innovation
Editor: Lean Karlo Santos Tolentino
ISBN: 979-8-89530-464-8

[5]Center for Artificial Intelligence and Nanoelectronics (CAIN), Integrated Research and Training Center, Technological University of the Philippines, Manila, The Philippines

Abstract

This study proposes a design of a fully automated harvester to streamline the operation of an indoor hydroponics system from growing plants that can be easily grown in this planting method, such as lettuce, spinach, and bok choy, to harvesting them with less human intervention. The harvester design uses the concept of a positioning-and-handling robotic arm that employs a combination of servo and stepper motors in its 3D-printed and mechanical parts. It also uses infrared sensors for object detection. Using an ATMega2560 microcontroller, the sensors, and the actuators are integrated to operate as a fully automated harvester. In harvesting the full-grown plants in two parallel hydroponic tubes with a 3-inch space between each slot, it took the harvester 3.955 minutes to harvest all eight plants. Meanwhile, to its parallel tube, which is 4 inches apart from the first tube, it took 5.63 minutes to complete the harvest.

Keywords: automated harvester, robotic arm, ATMega2560, positioning-and-handling, hydroponics planting

Introduction

Nutritious food is necessary for a healthy diet to improve an individual's overall health and well-being. Food insecurity and malnutrition go hand in hand as issues that the Philippines has been dealing with for years (FAO, IFAD, UNICEF, WFP and WHO, 2020). Food security is constantly affected by population growth and urbanization. An increasing fraction of the population is now experiencing hunger and malnourishment as agricultural fields are transformed to accommodate the changes brought by urbanization. Between 2017 and 2019, the Philippines had the highest number of food-insecure individuals in Southeast Asia, with a total of 59 million Filipinos experiencing a moderate to severe lack of continuous access to food (Baclig, 2022).

Through thorough investigation and research, this study aims to construct a Fully Automated Positioning and Handling System for Indoor Hydroponic

Crop Harvesting. This system will allow a sustainable and energy-efficient environment and hands-free farming of high-quality plants.

Contribution

Crops are traditionally and frequently harvested by hand. Manual harvesting is particularly common for crops with wide time periods for optimal maturation, crops with wide time periods for optimal maturation, or crops offered for direct consumption despite the fact that it is labor-intensive. Farmers turned out mostly complaining about their inability to find labor. The shortage in terms of labor has started to become a major issue for farmers, especially in states like California (Samangooei, SaSSi, & lack, 2016). As for the chefs, they were mostly looking for quality, of course, but also predictability and consistent quality. Thus, automation is sought out by most to lessen manual labor.

This present invention revolutionizes the traditional farming process through the integration of a fully automated harvesting system into the hydroponic system. The following emphasizes the relevance of the proposed automation of the harvesting process:

- A fully automated hydroponic harvester has the potential to streamline the harvesting process, reduce labor costs, and increase efficiency.
- This method doubles plant growth in half the time. This type of harvesting system could use sensors and cameras to gather data from the system, detect whether the plants are ready for harvest, and employ a robotic arm to cut and transport them to a processing area.

Related Works

Wang et al. developed an automatic harvester specifically intended for hydroponic lettuce. The design considered the physical and mechanical characteristics of the lettuce. Its focus is on analyzing the moisture content of the stem, root, and leaf, as well as the geometric properties, pulling force, and root-cutting force, to optimize the harvesting process of hydroponic lettuce. The cutting force exerted by the roots and the pulling force exerted by hydroponic lettuce were measured by shear and tensile experiments, respectively. Statistical analyses indicate that the leaves have the maximum moisture content, while the stems demonstrate the lowest level of toughness, making them more prone to breakage. The researchers additionally examined the root-cutting force of the hydroponic lettuce using various cutting speeds and cutting positions. The cutting position was influenced by the force exerted on the root during cutting. The tensile experiment yielded a mean force of 13.03 N as the pulling force of hydroponic lettuce.

The research offers sufficient theoretical justification for the design of an automated hydroponic lettuce harvester. The findings of this study have the potential to guide the development of future designs for automated harvesting equipment (Wang et al., 2020).

Pham et al. (2020) introduce a robotic harvesting system that utilizes computer vision technology in hydroponic farms to collect lettuce. The researchers would estimate the location of the harvested lettuce based on the positions of the holes in the hydroponic system. The positions of holes concealed by lettuces are approximated by utilizing Hough transforms to identify uncovered holes and the borders of tubes. The harvesting task employs a robot manipulator and a gripper controlled by a servo mechanism. The final findings indicate that the system, with an average location error of 0.83 mm (in everyday scenarios) and a maximum error of 9.62 mm, can effectively carry out harvesting in real-world settings.

Similarly, Liu et al. (2023) developed a novel mechanical harvesting mechanism for hydroponic leafy vegetables cultivated in pipelines to address the inefficiencies and labor-intensive nature of manual harvesting. The research specifically targets hydroponic Chinese kale, a prevalent variety in South China, and proposes an innovative grabbing mechanism with double-pivot rotation cross fingers. This mechanism is designed to envelop the grabbable area of the vegetable, center its stalk in the planting hole, and remove it without collision with the extended leaves. Laboratory experiments demonstrated that the initial inclination angle of the vegetable stalk significantly impacts the grasping success rate, with optimal results achieved at angles between 85° and 95° and finger deflection speeds of 40° s-1 to 60° s-1, achieving a success rate exceeding 95%. This study provides foundational insights for the development of automated harvesting equipment, highlighting the critical role of precise positioning and orientation in enhancing the efficiency and productivity of hydroponic vegetable cultivation systems.

Zhang et al. (2016) present a wolfberry harvesting robot's design and control system, addressing manual harvesting challenges, such as slow speed, low efficiency, and poor sanitary conditions. The proposed robot consists of a self-propelled automated platform with an articulating arm, equipped with cameras, sensors, and a specialized picking manipulator. The picking mechanism utilizes two silicone wheels with spiral silicon tubes that simulate hand-picking through relative motion, allowing for gentle fruit removal without damaging leaves or immature fruit. The robot's control system employs pinhole imaging technology for fruit identification and location and a fuzzy-PID control method to enhance the system's dynamic and static performance.

Raja et al. (2022) discuss an integrated robot system for agricultural harvesting designed to be affordable for small and medium-scale farmers. The system combines a mobile automated guided vehicle (AGV) with a Cartesian and articulated robot arm for crop detection and picking. It uses deep learning, specifically the YOLO-V3 algorithm, for crop detection through computer vision. The robot was tested on carrot and cantaloupe crops, achieving detection accuracies of 93% and 95% respectively. For navigation, the system uses GPS waypoint planning to cover the farming area. The author notes that while the robot's harvesting speed of about 18 seconds per crop is slower than manual harvesting, it can work continuously without tiring.

Using inverse kinematics, Lauguico and Bandala (2019) propose a framework for optimizing a robotic arm to harvest crops in vertical farming setups. The researchers implemented the system using MATLAB simulation and a Universal Robot (UR) hardware setup. They developed an approach involving initializing waypoints, processing them through inverse kinematics, and directing the robotic arm's movements to target crop locations on a wall garden. The system aims to automate crop harvesting in vertical farms by precisely positioning the robotic arm's end-effector to grip and pull crops. Testing involved four different waypoint sequences, with results showing an average 85.42% success rate in gripping and pulling test objects representing crops.

Meanwhile, Zhang et al. (2016) presented a novel design for a robot manipulator specifically aimed at harvesting strawberries in ridge-culture systems using Generalized Function (GF) set theory to design a 4-PPPR parallel robot manipulator with three translational degrees of freedom (DOF) and one rotational DOF. This design approach allows for a "top-down" logic in determining the manipulator's configuration based on the desired features of the moving platform. The researcher highlights the advantages of parallel mechanisms in agricultural robotics, including compact structure, low inertia, high rigidity, and accuracy. These are particularly beneficial in situations requiring precise movements in limited spaces.

Moreover, Bao et al. (2015) developed a flexible pneumatic bending joint using the FPA and modeled its bending angle and output force in relation to air pressure. Based on this joint, they designed a three-fingered end-effector specifically for grasping and picking fruits like apples. The end-effector incorporates a rotary shear mechanism for cutting fruit stems. Experiments were conducted to validate the mathematical models and test the grasping performance of apples.

Hui et al. (n.d.) developed an indoor hydroponics robot system for seeding and logistics in small-scale urban farming environments. The system consists of three main components: a seeding robot with a custom end-effector for planting small seeds, mobile robots for tray logistics, and a management system for scheduling and operation. The seeding robot uses a 6-DOF arm with a modular cassette and air-assisted needle array to accurately place seeds as small as 1-2 mm in diameter. Computer vision and the Iterative Closest Point algorithm ensure precise seed placement. For tray logistics, small mobile robots navigate tight spaces using AprilTags for docking and a centralized server for collision avoidance. The system includes a web-based interface for task scheduling and monitoring. The authors note that while the system is adaptable to various farm sizes, limitations include seed shape restrictions and lack of harvesting capabilities, suggesting areas for future development.

Peter P. Ling et al. (2004) presented a study on developing an automated system for harvesting tomatoes. The robotic system developed includes a sensing unit and an end-effector integrated with a commercial robotic manipulator. The sensing unit utilizes image processing algorithms to identify and locate mature tomatoes, even those partially hidden by foliage. The end-effector, equipped with a four-finger prosthetic hand and an embedded controller, is designed to pick, hold, and place the tomatoes without damaging the plant. This system demonstrated high success rates in laboratory and commercial greenhouse environments, achieving over 95% accuracy in fruit sensing and over 85% in fruit picking. The research contributes significantly to the field of agricultural automation by addressing the challenges of selective harvesting and ensuring continuous production of high-quality fruits.

Tarrío et al. (n.d.) focused on the complex task of harvesting small fruits in bunches, such as strawberries, by developing a 3D stereoscopic vision system. This innovative approach, combining stereoscopic cameras with structured lighting, enables the precise identification and localization of individual fruits within clusters, addressing a significant challenge in automated fruit harvesting.

Concurrently, Birrell et al. (2019) tackled the difficulties associated with iceberg lettuce harvesting through their Vegebot system. By integrating advanced computer vision techniques with a specialized end-effector, they achieved high success rates in lettuce localization and classification, demonstrating the potential for robotics in delicate crop harvesting.

Expanding the scope to urban environments, Moraitis et al. (2022) developed the CityVeg system for automated urban garden management. This compact robotic platform incorporates precision irrigation, plant identification, and cloud-based data management, addressing the growing interest in urban agriculture and the need for efficient small-scale farming solutions.

Methodology

The harvesting automation using the robotic arm will use the Arduino MEGA board with an ATMega2560 microcontroller. The indoor hydroponic system's operation is divided into Growing (Hydroponics), and Harvesting. In the planting process, the germinated seeds are placed manually inside the chamber with a hydroponic foam in a cup along with the germinated seed, and they are placed in one of the holes of the hydroponic tubes. For 3 to 4 weeks, the germinated seeds will develop into full-grown plants with the help of the balanced and sufficient nutrients in the solution that will be measured and regulated automatically by the system. In the process of Growing, the crops are monitored using image processing. Also, when the growing period is completed and the plants are detected as Ready-to-Harvest, they will be harvested by the robotic arm, gripping the cup, and placing them into the harvest outlet located inside the chamber.

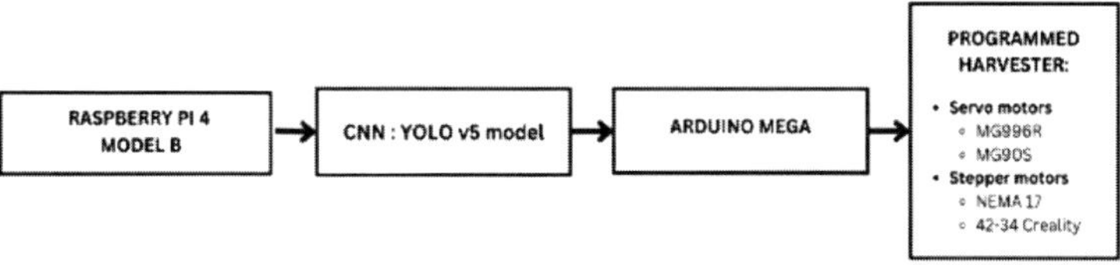

Figure 1. Automated harvesting system block diagram.

When the plants are Ready-to-Harvest, the Arduino MEGA connected to the robotic arm receives a certain signal from the image processing system, harvesting the full-grown plants as shown in Figure 1.

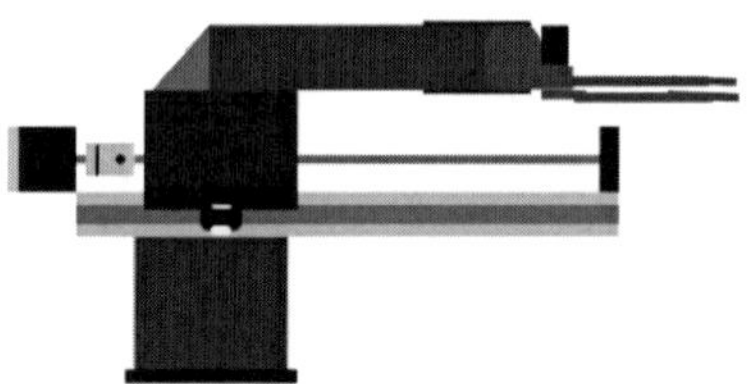

Figure 2. Automated harvester side view.

The automated harvester employs a gripper as its end-effector to harvest the plants. The gripper is based on an open-source pick-and-place 6 DOF robotic arm design. In contrast, its body was designed by researchers depending on the indoor hydroponic system's structure, as seen in Figure 2. The robotic arm is attached to a combination of vertical and horizontal gantry. The body of the arm is constructed using 1 kg of PETG filament through 3D printing technology. To pivot the joints to a desired angle, two types of servo motors are used –MG996R for the arm part and MG90s for the gripper part. The programmed sequence of motor movements will perform the harvesting feature of the system. Furthermore, to suffice the required scale of functionality of the dimensions of the chamber, a vertical and horizontal gantry aids the movement of a single robotic arm to accommodate the number of plants on all four layers.

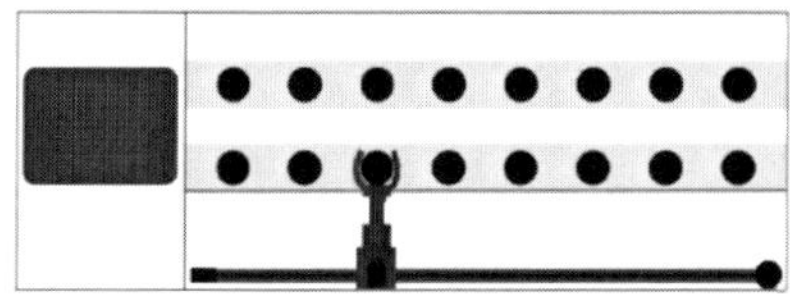

Figure 3. Automated harvester top view with the two parallel hydroponic tubes.

In the vertical gantry, a GT2 pulley belt and a stepper motor elevate the horizontal gantry to a certain layer of hydroponics tube. In turn, the horizontal gantry also uses the timing belt system to aid the robotic arm in reaching the plants shown in Figure 3. The movement of the stepper motors is also programmed so that the harvester can reach an accurate position to grip the mouth of the cup with a full-grown plant. Figure 4 shows the process of the automated harvester's operation.

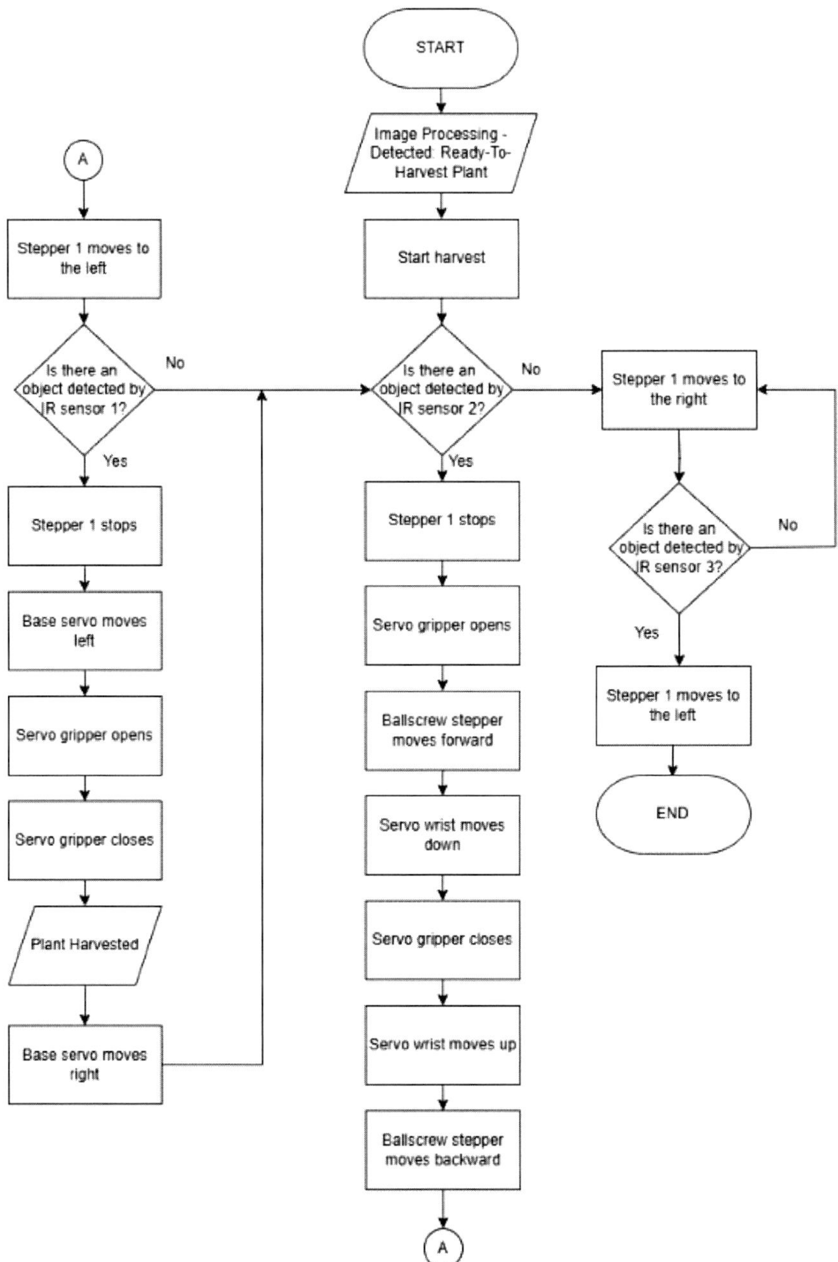

Figure 4. The automated harvesting system's flowchart.

Results and Discussions

The implementation of the automated harvester design shown in Figure 5 has successfully operated as a positioning-and-handling robotic arm and harvested the load of the cup with a full-grown plant.

Figure 5. The actual automated harvester.

Referring to the image shown in Figure 6, the automated harvester can grip the cup holding the plant and lift and hold its weight while it is being transported to the basket where the harvested plants are placed. The time taken to harvest the eight full-grown plants in one hydroponic tube is 3.955 minutes as shown in Figure 7, while to its parallel tube, it is 5.63 minutes shown in Figure 8.

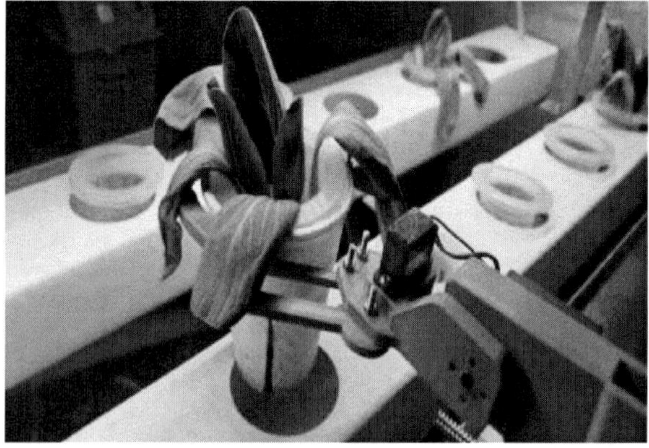

Figure 6. The automated harvester picking a full-grown bok choy.

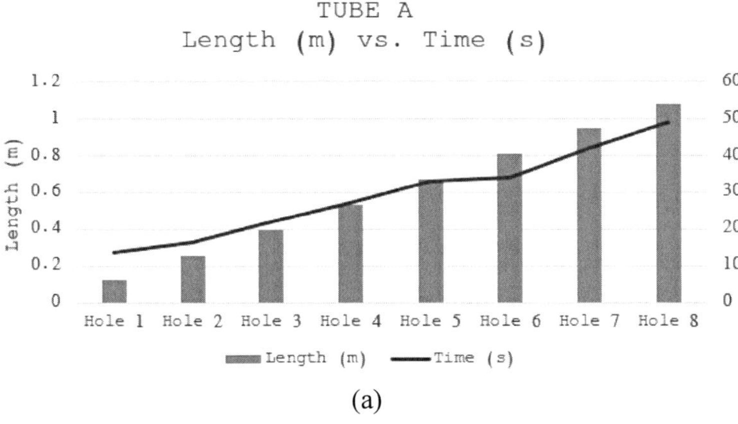

(a)

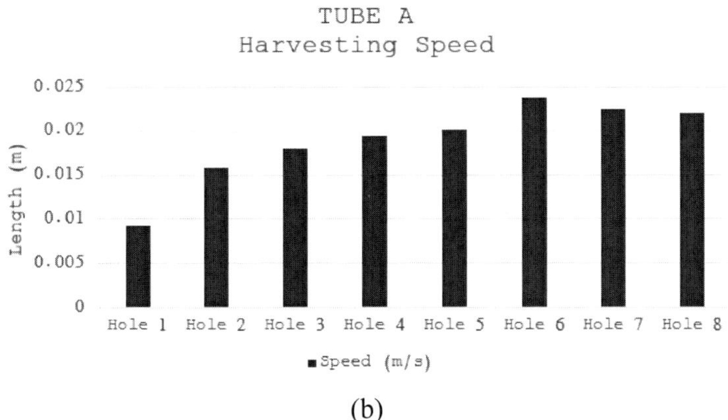

(b)

Figure 7. Harvesting speed through-out the Tube A: (a) length vs. time (b) harvesting speed.

For an indoor hydroponic system with two parallel hydroponic tubes with eight slots each and with a distance of 4 inches in between, the total time taken for this design of automated harvester is 10.18 minutes. The above figures show that an automated harvester using a NEMA 17 stepper motor experienced inconsistent harvesting speed despite uniform hole spacing in the tube due to several factors. Motor torque fluctuations at different speeds can cause variations in performance, especially if the load is not perfectly balanced. Microstepping, while providing smoother motion, can introduce slight positional inaccuracies that accumulate over time. Mechanical factors like friction and vibrations can also affect consistency. Additionally, variations in power supply voltage or current can impact motor performance.

Lastly, environmental conditions such as temperature and humidity may also influence the system's behavior.

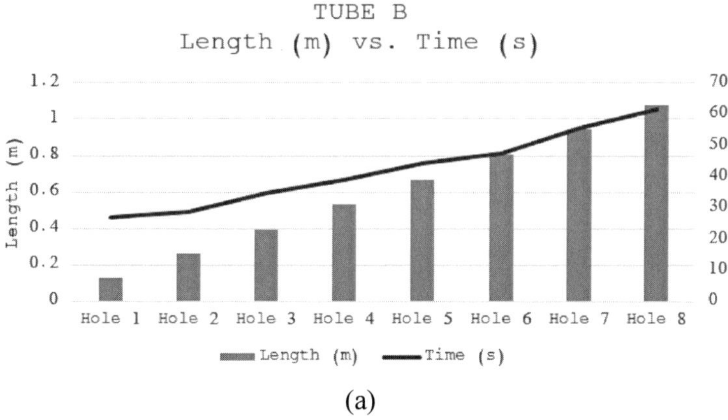

(a)

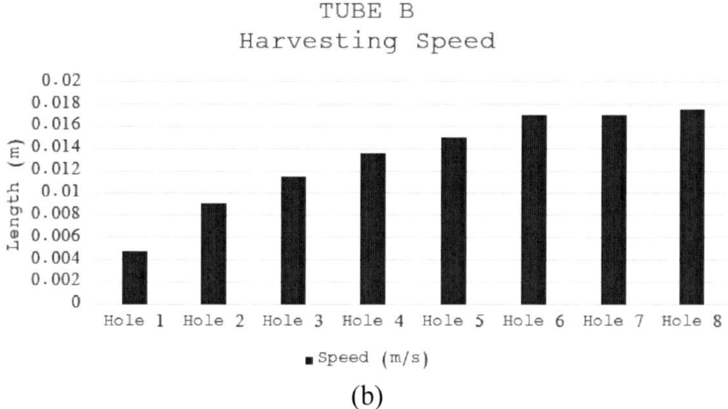

(b)

Figure 8. Harvesting speed throughout the Tube B: (a) length vs. time (b) harvesting speed.

Conclusion

In conclusion, this design of the automated harvester effectively harvests 16 full-grown plants, such as lettuce, spinach, and bok choy, in an indoor hydroponic system within 10.18 minutes. With the help of a fully automated harvester, there is less manual labor in harvesting plants, which can reduce labor costs and pave the way for the reality of fully automated farming systems.

Integrating this automated harvesting system into a monitoring system that includes image processing, which can determine whether the plants are ready to harvest, and an automated indoor hydroponic system that efficiently produces full-grown plants with minimal manual labor has proven to be feasible and effective.

Future Works

The researchers of this study endorse the following to enhance and make further advancements in the automated harvesting system:

1. The plants grown in hydroponics also had roots. To make matters more convenient, a system to cut the roots of the plants before harvesting is recommended.
2. Once the plant is harvested, the slot allotted for it will be empty. To lessen human intervention, adding a planting feature to the arm is recommended.
3. To make the automated harvesting system more intelligent and efficient, the harvesting mechanism will be improved by integrating image processing in place of infrared sensors.

Disclaimer

None.

References

Baclig, C. E. (2022). World Food Day 2022: Rising costs keep millions in PH away from healthy diets. PIDS – Philippine Institute for Development Studies.

Bao, G., Yao, P., Cai, S., Ying, S., & Yang, Q. (2015). Flexible pneumatic end-effector for Agricultural Robot: Design & Experiment. In *2015 IEEE International Conference on Robotics and Biomimetics (ROBIO). https://doi.org/10.1109/robio.2015.7419096.*

Birrell, S., Hughes, J., Cai, J. Y., & Iida, F. (2019). A field-tested robotic harvesting system for Iceberg Lettuce. *Journal of Field Robotics,* 37(2), 225–245. https://doi.org/10.1002/rob.21888.

FAO, IFAD, UNICEF, WFP and WHO. (2020). The State of Food Security and Nutrition in the World 2020: Transforming food systems for affordable healthy diets. Rome: FAO. https://doi.org/10.4060/ca9692en.

Hui, H. K., Lee, J. H. M., Sum, A. K. W., & Lau, D. (n.d.). *Indoor Hydroponics Robot System for Automated Seeding and Logistics.* Retrieved from http://www.cse.cuhk.edu.hk/~jlee/publ/24/icca24VMOrobots.pdf.

Lauguico, S. C., Concepcion, R. S., Macasaet, D. D., Alejandrino, J. D., Bandala, A. A., & Dadios, E. P. (2019). Implementation of inverse kinematics for crop-harvesting robotic arm in vertical farming. In *2019, IEEE International Conference on Cybernetics and Intelligent Systems (CIS) and IEEE Conference on Robotics, Automation and Mechatronics (RAM).* https://doi.org/10.1109/cis-ram47153.2019.9095774.

Ling, P. P., Ehsani, R., Ting, K. C., Chi, Y.-T., Ramalingam, N., Klingman, M. H., & Draper, C. (2004). Sensing and end-effector for a Robotic Tomato Harvester. *Proceedings of the 2004 ASAE Annual International Meeting,* Ottawa, Canada, August 1 - 4, 2004. https://doi.org/10.13031/2013.16727.

Liu, Y., Xia, H., Feng, J., Jiang, L., Li, L., Dong, Z., Zhao, K., & Zhang, J. (2023). An enveloping, centering, and grabbing mechanism for harvesting hydroponic leafy vegetables cultivated in the pipeline. *Agronomy*, 13(476). https://doi.org/10.3390/agronomy13020476.

Moraitis, M., Vaiopoulos, K., & Balafoutis, A. T. (2022). Design and implementation of an urban farming robot. *Micromachines,* 13(2), 250. https://doi.org/10.3390/mi13020250.

Raja, V., Bhaskaran, B., Nagaraj, K. K., Sampathkumar, J. G., & Senthilkumar, S. R. (2022). Agricultural harvesting using integrated robot system. *Indonesian Journal of Electrical Engineering and Computer Science,* 25(1), 152-158. https://doi.org/10.11591/ijeecs.v25.i1.pp152-158.

Samangooei, M., SaSSi, P., & Iack, A. (2016). Soil-less systems vs. soil-based systems for cultivating edible plants on buildings in relation to the contribution towards sustainable cities. *Future of Food: Journal on Food, Agriculture and Society*, 4(2), 24–39.

Tarrío, P., Bernardos, A. M., Casar, J. R., & Besada, J. A. (n.d.). A harvesting robot for small fruit in bunches based on 3-D Stereoscopic Vision. *Computers in Agriculture and Natural Resources,* 23-25 July 2006, Orlando, Florida. https://doi.org/10.13031/2013.21885.

Viet-Cuong, P., Hoang-Giap, N., Thanh-Khang, D., & Gia-Bao, H. (2020, November 11). A Computer Vision Based Robotic Harvesting System for Lettuce. *International Journal of Mechanical Engineering and Robotics Research.* Retrieved from http://www.ijmerr.com/uploadfile/2020/1016/20201016113846679.pdf.

Wang, W., Ma, Y., Fu, L., Cui, Y., & Majeed, Y. (2020). Physical and mechanical properties of hydroponic lettuce for automatic harvesting. *Information Processing in Agriculture*. https://doi.org/10.1016/j.inpa.2020.

Zhang, K., Zhang, T., & Zhang, D. (2016). Synthesis design of a robot manipulator for strawberry harvesting in ridge-culture. In *2016 Asia-Pacific Conference on Intelligent Robot Systems (ACIRS).* https://doi.org/10.1109/acirs.2016.7556198.

Zhang, X., Zhao, Z., Wei, S., Du, G., Ji, Q., Zhang, L., & Wei, H. (2016). Study on the design and control system for Wolfberry Harvesting Robot. In *2016 Chinese Control and Decision Conference (CCDC).* https://doi.org/10.1109/ccdc.2016.7532069.

Chapter 3

CYNOSENS: An IoT-Based Harmful Algal Bloom Prediction System Using Random Forest Regression Through Water Quality Parameters

Vincent Delos Reyes[1]
Ariel Fernando Jr.[1]
Verlyn Ann Jane Onrada[1]
Louis Edward Sabordo[1]
Carl Vincent Tioco[1]
and Gilfred Allen Madrigal[1,2,*]

[1]Department of Electronics Engineering, Technological University of the Philippines, Manila, The Philippines
[2]Center for Engineering Design, Fabrication, and Innovation, College of Engineering, Technological University of the Philippines, Manila, The Philippines

Abstract

One of the various economic uses of different bodies of water is aquaculture; however, one of the major problems it deals with nowadays is fish kills that are caused by fluctuations in the natural environment that lead to eutrophication — a process that results in the accumulation of harmful algal blooms (HABs). HABs are dangerous for aquatic systems as they slow the growth of fish and other aquatic animals. Thereby, the study aimed to develop a standalone monitoring device that is efficient and a real-time system to detect the presence of algal blooms and predict when they will become harmful to aquatic animals. Specifically, the

[*] Corresponding Author's Email: gilfredallen_madrigal@tup.edu.ph

In: Aquaponics: Sustainable Farming, Ecology and Innovation
Editor: Lean Karlo Santos Tolentino
ISBN: 979-8-89530-464-8

monitoring buoy is equipped with six water quality sensors for the following parameters: pH, Temperature, Conductivity, Dissolved Oxygen, Nitrate, and phosphate, utilizing Random Forest Regression as a predictive algorithm and the Internet of Things. The monitoring device is connected to the web application via a wireless network, and the prediction function of the device is based on the time it was brought into the water to undergo a quality test. The web app displays a warning that alerts users to HABs and whether specific measured values are already detrimental to the fish. The system triggered at a 21 µg/L level of algae in the water and detected harmful algae content greater than 50 µg/L with 91.48% prediction model accuracy, which is beneficial as it can determine an extensive range of HAB levels in the water. Results demonstrated that HAB monitoring buoys could rapidly and efficiently supplement manual sampling in at-risk water bodies.

Keywords: algal bloom, aquaculture, IoT

Introduction

The Philippines is fortunate to have abundant natural resources; one endowed resource is water. The aquatic biome of the country is classified into two categories: freshwater and marine regions. Rivers, lakes, and groundwater are examples of inland freshwater, whereas bay, coastal, and oceanic waters are examples of marine water. These bodies of water are vital to the country's economic development since they support cultivating various crops that meet an individual's demands. One economic use of diverse bodies of water in the Philippines is aquaculture; it involves breeding, raising, and harvesting various fish species and aquatic plants. Likewise, rapid aquaculture growth has arisen since the discovery that different species of fish, including those not native to its water, can be grown in controlled environments in the lake. Nowadays, despite its importance, due to rapid population growth, urbanization, industrialization, and a lack of access to safe and sufficient water, the quality of Philippine waters is diminishing, a serious problem facing the country today. Wastewater discharged from domestic, industrial, and agricultural runoff contains detergents, fertilizers, heavy metals, chemicals, oils, and solid waste materials, seriously damaging the recipient water bodies. Each of these pollutants has distinctive toxic repercussions that affect human livelihood, resulting in economic costs.

In line with the detrimental activities that affect the aquaculture of the Philippines, several problems are being encountered by fishermen in different regions, and the most common problem is fish kill, which occurs because of human activities and natural causes. Fishkill, defined by the Bureau of Fisheries and Aquatic Resources (BFAR), is the "massive death of fish stocks in a specific area as a result of unfavorable water quality parameters of a given aquatic environment that are intolerable or toxic to the fish stocks." Additionally, the body of water comprises several parameters, namely pH level, turbidity, conductivity, temperature, salinity, and many others that are vital in perceiving and understanding the quality of the given water. Each parameter has a different degree of importance depending on the purpose of the study that has to be done in the water.

The formation of harmful algal blooms (HAB) is one of the natural causes, and when not treated immediately, it could lead to fish death. HAB is the rapid enrichment of algae or cyanobacteria that occurs on the surface of lakes and other bodies of water when the nutrients on which algae depend are overabundance in water. Thus, the unfavorable water quality parameters cause the algae to bloom in the water. Moreover, algal blooms are harmful because they reduce fish and other aquatic life's ability to find food, block sunlight from underwater plants, and consume oxygen — insufficient oxygen makes it impossible for aquatic life to survive. On this note, it is highly relevant to predict the bloom of harmful algae to protect the aquatic life and for the fisherfolk to know when to take necessary action to prevent it.

There are six water quality parameters responsible for harmful algal bloom occurrence in water, namely: pH level, nitrate, phosphorus, temperature, electrical conductivity, and dissolved oxygen. Similarly, low levels of dissolved oxygen, ammonia, nitrite, or hydrogen sulfide in shrimp and fish cause stress and illness. Ammonia and hydrogen sulfide toxicity can be increased by temperature and pH abnormalities—moreover, excess nitrogen and phosphorus result in algae growth (Prananingtyas et al., 2019). Hence, having appropriate amounts of these water quality indicators plays an important role in preventing harmful algal formation.

With the advancement of technology, various tools and sensors were developed to monitor water quality parameters. These include the use of electrochemical and optical sensors, among many others. Algal bloom monitoring is done by water sampling, and the parameters are being studied in laboratories. In addition, a water station buoy was developed to remotely monitor the water parameters to determine the quality of a given water.

However, most of these devices are imported from other countries and are particularly expensive.

The development of computing capabilities has brought modern techniques to predict the harmful algal bloom in water. An in-depth literature review encompassing different applied machine learning techniques models for predicting harmful algal bloom, artificial neural network (ANN), and support vector machine (SVM) are two commonly used. Conversely, the experimental results show that the improved backpropagation (BP) algorithm and SVM perform better than (generalized regression neural network) GRNN methods. Even so, analysts still recommend that these algorithms be further improved.

In this study, the machine learning algorithm employed in a certain data set is Random Forest Regression. Random Forest is a versatile and easy-to-use machine learning algorithm that consistently delivers good results without hyperparameter adjustment. It is also one of the most extensively used algorithms due to its simplicity and can be used for classification and regression problems. The "forest" it generates is a collection of decision trees trained using the "bagging" method. Bagging techniques are based on the concept that combining learning models enhances overall performance.

This has encouraged the researchers to develop a standalone prediction system that predicts the presence of HABs in bodies of water by monitoring water quality indices. The device used water sensors for dissolved oxygen, water temperature, conductivity, nitrate, phosphate, and pH level. The prediction system was programmed using Random Forest Regression, and the data collected by the device was displayed in a web application.

Review of Related Literature and Studies

This chapter synthesizes related literature and studies so that researchers can gain an in-depth understanding of the relevant information. Hence, this chapter establishes the foundations of the concepts used and how they will be used in the context of the study.

Aquaculture

Aquaculture is the controlled cultivation of aquatic organisms such as fishponds and fish pens. According to Demetillo et al. (2019), aquaculture covers almost 50% of fish on the planet that is being consumed by humans and is expected to increase in the following years. Lorenzo et al. (2019) also said that the aquaculture sector around the globe has continuously grown in the past 40 years, though comparatively, the aquaculture industry in some countries has grown more rapidly than others. In connection with the large contribution of aquaculture to the world's economy, various factors are being considered, such as its contribution to pollution, disease management, antibiotic use, food safety, access to water, etc.

State of Aquaculture in the Philippines

The Philippines archipelago is surrounded by different bodies of water. With more than 7000 islands, various bodies of water are also being utilized for aquaculture (Anderson et al., 2019). Aquaculture is one of the Philippines' main food sources; it has provided livelihood to different locals residing near lakes and seaside. However, given that the country is highly dependent on fish products, aquaculture in the Philippines has been declining because of the decrease in seaweed production (Nadarajah & Flaaten, 2017). According to the Philippine Statistics Authority, the Philippines' aquaculture provided 2,322.91 metric tons of fish products in 2020 and 2,246.32 metric tons in 2021. These numbers show a decrease of 3.3% in the yearly production of fish products in the Philippines (Borlaza et al., 2023).

Fish Kills in the Philippines

Alongside the benefits and different uses of water bodies in the Philippines and aquaculture, problems usually arise, especially fish kills, which can be attributed to different causes. In Taal Lake, fish kills were caused by isopod infestation, sulfur upwelling, abnormal temperatures, anomalies in dissolved oxygen levels, and other physicochemical parameters (White, 2017). On the other side of CALABARZON, algal blooms occur periodically in Laguna Lake. Cyanobacteria cause HABs or Harmful Algal Blooms that consume oxygen in the lake, leading to the death of fish caused by oxygen depletion (Mapa, 2022).

In May 2019, a fish kill incident occurred in Biñan and Pila, Laguna, as thousands of dead tilapias and bangus were scattered in the fishing zone. An estimated 500,000 fish have died while the remaining produce foul odor, which is no longer profitable. Ireneo G. Bongco, Senior Science Research Specialist at Laguna Lake Development Authority (LLDA), said that the fish kill was caused by a green tide (red tide in saltwater) or algal blooms, where the fish suffered from asphyxia due to deficiency in dissolved oxygen (DO) level concentration brought by the competing thick layer of green tide in the lake surface. These fish kills caused by oxygen deficiency always damage the fishermen's livelihood around the lake (Torres, 2020).

Additionally, on April 18, 2022, the public in Bolinao, Pangasinan, was advised to refrain from consuming all types of shellfish harvested from coastal waters due to the presence of algae that produces the red tide toxin (Mendoza et al., 2019).

Table 1. Synthesis for fish kills in the Philippines

Author	Year	Title	Relevant Findings	Relationship to the Study
Anderson J., Asche F., and Garlock T.	2019	Economics of Aquaculture Policy and Regulation	Aquaculture has made a big contribution to the world's economy and its contribution is continuously growing.	The study focuses on factors affecting the habitat of fish because the occurrence of fish kills has an impact on the economy.
Nadarajah S., and Flaaten O.	2017	Global Aquaculture growth and institutional quality	The global Aquaculture industry has been continuously growing.	Monitoring fish habitat is vital to the aquaculture industry.
Mapa D.	(2022).	Fisheries Situation Report for Major Species	Aquaculture in the The Philippines has provided an annual amount of approximately 2000 metric tons of fish, but the annual harvest has a decline rate of 3%.	The decline in the amount of fish harvested could be the result of fish kills occurring and poor habitat management.
Tiongson K., Tamayo-Zafar alla M.	2018	Water Quality and Seasonal Dynamics of Phytoplankton and Zooplankton in the West Bay of Laguna de Bay, Philippines	Harmful Algal Blooms consume dissolved oxygen in water.	Competition to Dissolved Oxygen leads to fish kill. Monitoring Dissolved Oxygen levels may prevent fish kills.
Visperas E.	2022	Red Tide Warning Up in Pangasinan	Due to the presence of algae that creates the red tide toxin, shellfish taken near ocean areas are unsafe.	The possible location of the deployment of the device

Algal Blooms

Harmful algal blooms (HABs), sometimes known as "red tides," are a problem in almost all coastal nations worldwide. These occurrences are brought on by blooms of minuscule algae. Some of these algae are harmful, and exposure to them can cause disease and even death in people, marine mammals, seabirds, fish, and other aquatic life. Toxins are often spread through the food chain. The direct discharge of harmful substances can occasionally kill marine life. Due to the vast amount of algae collected, non-toxic HABs are often harmful to ecosystems, fish stocks, and recreational facilities. The term "HAB" also refers to non-toxic macroalgal blooms (seaweeds), which can have significant ecological effects such as displacing native species, changing habitats, and depleting bottom-water oxygen (Mendoza et al., 2019).

On the other hand, Freshwater Microcystis is a global genus and has developed algal blooms in Laguna de Bay since the early 1970s (Tiongson & Tamayo-Zafaralla, 2018). Algal blooms are a natural occurrence. However, it is one of the nuisances in coastal areas (Anderson, 2009). Blooms are dense clumps of cyanobacterial cells that form in marine, brackish, and freshwater environments and are usually noticeable by the visible discoloration they cause in the water (Espaldon et al., 2005). Algal blooms caused by high fertilizer concentrations also lead to huge changes in dissolved oxygen (DO) concentrations. Large volumes of dissolved oxygen are present in surface waters throughout the daytime during blooms due to photosynthesis. However, at night, hypoxia or anoxia brought on by algal cell respiration could result in fish deaths (U.S. EPA, 2023).

Eutrophication

Water eutrophication is one of the environmental issues we faced until now. Most freshwater lakes encounter water quality problems and ecological imbalances caused by anthropogenic activities (Heisler et al., 2008). Specifically, nitrogen and phosphorus eutrophication in marine ecosystems. It has a negative impact on food security, ecosystem health, and the economy through disruptions in fisheries. Uncontrolled levels of nitrogen and phosphorus cause algae blooms, anoxic conditions, and ocean acidification, leading to dead zones and fish kills (D'Silva et al., 2012). Moving forward, lakes grow more fertile and shallow due to eutrophication, the process of lakes receiving phosphorus and nitrogen as nutrients and silt from the nearby watershed. The more nutrient-rich a lake is, the more water organisms it can support (D'Silva et al., 2012).

Different factors, such as agricultural, municipal, and industrial activities, have significantly increased aquatic nitrogen and phosphorus pollution. This resulted in the water becoming fertile, known as eutrophication, thus endangering water quality and biotic integrity from headwater streams to coastal locations. The excessive nutrients in the water contribute to oxygen depletion, ecosystem diversity, and eutrophication (Ngatia et al., 2019) Consequently, Hypoxic "dead zones" reduce fish and shellfish production, harmful algal blooms cause taste and odor problems and threaten the safety of drinking water and aquatic food supplies; greenhouse gas release stimulation; and degradation of cultural and social values of these waters are all problems caused by eutrophication (RMBEL, 2018).

Monitoring of Algal Blooms

Throughout the lifespan of a desalination plant, water quality monitoring is essential for identifying occurrences that result in poor water quality, such as harmful algal blooms (HABs). The organic and solids load in the seawater feed to be treated at a desalination plant might significantly rise due to HABs. The crucial for process control and contractual objectives is to monitor the water quality of the intake feedwater to optimize pretreatment procedures in reaction to a decline in feedwater quality brought on by situations such as HAB to sustain both production and quality targets (Oracle Philippines, n.d.).

The suggested model for chlorophyll-a concentration has three primary novelties. In order to explicitly address the complex spatio-temporal dependence in chlorophyll-a concentration on a monthly scale, the MHMM can be used as a framework using the latent (or hidden) classes, which correspond to the mixture components for each observation and are supposed to follow a Markov process. A multivariate hidden Markov model (MHMM) based on gamma mixture distribution was suggested to achieve these goals and construct a decision-support framework for algal bloom at several stations in the lower Nakdong River. The study intends to investigate the spatio-temporal dynamics of phytoplankton dispersion with chlorophyll-a concentration and perhaps cluster them using a limited number of latent states, which can be connected to other water quality measures (Wilhelm, 2009).

Prediction of Algal Blooms

Each year, harmful algal blooms cause environmental harm, financial losses, and outbreaks of diseases. Since there is yet no fundamental resolution to this concern, improving early warnings (predictions) is the best approach to mitigate the effects of algal blooms. Although numerous variable data sources are needed for the study, current physical prediction models struggle to provide a precise coefficient demonstrating the relationship between each aspect when predicting algal blooms. High time and financial costs come along with these constraints. While this is happening, artificial intelligence and deep learning techniques are becoming more and more prevalent in scientific research. As a result, more and more long short-term memory (LSTM) model applications to environmental research challenges are being made because the LSTM model performs well for time-series data prediction. Few studies, particularly in South Korea, where annual algal blooms occur, have used deep learning models like LSTM to predict algal blooms. As a result, the researchers used the LSTM model to predict algal blooms in South Korea's four main rivers. Researchers used regression analysis and deep learning approaches to make short-term (one-week) predictions on a recently created water quality and quantity dataset derived from 16 damned pools on the rivers. Chlorophyll-a, a well-known proxy for algal activity, was predicted using three deep learning models (multilayer perceptron, MLP; recurrent neural network, RNN; and long short-term memory, LSTM). The outcomes were contrasted with actual data based on the root mean square error (RSME) and OLS (ordinary least square) regression analysis. The LSTM model generates the highest predictive rate for harmful algae outbreaks, and all deep-learning models outperformed the OLS regression analysis. The findings demonstrated the potential of LSTM and deep learning for algal bloom prediction (Wurtsbaugh et al., 2019).

The researchers developed several machine-learning-based predictors, including AdaBoost (Adaptive Boosting), ANN (Artificial Neural Network), GBDT (Gradient Boosting Decision Tree), KNN (K-nearest Neighbor), and SVM (Support Vector Machine) for model selection in order to improve the accuracy of algal bloom predictions. Moreover, the researchers used an extensive testing procedure to evaluate the performance of all feature subsets under these predictors in order to identify the key environmental variables that influence prediction accuracy. Researchers discovered that the GBDT performs better on both datasets when using particular input combinations after empirical training (Yu et al., 2021).

Table 2. Synthesis for prediction of algal blooms

Author	Year	Title	Relevant Findings	Relationship to the Study
Ngatia, L., Grace III, J. M., Moriasi, D., and Taylor, R.	2018	Nitrogen and Phosphorus Eutrophication in Marine Ecosystems	Eutrophication is a major problem in aquatic ecosystems driven primarily by nitrogen and phosphorus.	Nitrogen and phosphorus are some of the parameters that the study focuses on.
Tsang, Y. F., and Kwon, H.-H.	2020	Stochastic Modeling of Chlorophyll-a for Probabilistic Assessment and 2 Monitoring of Algae Blooms in the Lower Nakdong River, South Korea	Phytoplankton abundance is generally indicated by chlorophyll-a concentration.	Chlorophyll-a is also present in algal blooms aside from cyanobacteria. Factors affecting the growth of cyanobacteria also affect the increase in chlorophyll-a.
Lee, S., and Lee, D.	2018	Improved Prediction of Harmful Algal Blooms in Four Major South Korea Rivers Using Deep Learning Models	The annual occurrence of harmful algal blooms causes environmental damage, economic losses, and disease epidemics.	Both studies predict the bloom of algae. However, the present study uses a Random Forest Regression algorithm.
Anderson, D. M.; Boerlage, S. F. E. and Dixon, M.B.	2017	Harmful Algal Blooms (HABs) and Desalination: A Guide to Impacts, Monitoring and Management.	Given the patterns of population, agriculture, development, and climate worldwide, an increase in hazardous algal occurrences is unavoidable.	The study focuses on monitoring and management of Harmful Algal Blooms.
Yu, P., Gao, R., Zhang, D., and Liu, Z.P.	2021	Predicting coastal algal blooms with environmental factors by machine learning methods	A machine learning-based approach that uses environmental factors to forecast the presence of algal blooms	The significance of each feature for the execution of the prediction indicates key elements for the occurrence of dangerous algal blooms.

Water Quality

One of the most important aspects of human life is water. It is essential to plants' and animals' survival and long-term viability. However, today, it is being widely exploited by dumping human garbage, industrial waste, and chemical waste into the water bodies, contaminating the water with harmful compounds. Water scarcity and pollution have been severe problems in many parts of the world. There is a lot of emphasis on industrialization and urbanization in many emerging countries, like India. This shows that water usage has increased, and quality has degraded (Kim et al., 2020).

Table 3. Synthesis for water quality

Author	Year	Title	Relevant Findings	Relationship to the Study
G. S. Menon, M. V. Ramesh, and P. Divya	2017	A low-cost wireless sensor network for water quality monitoring in natural water bodies	Evaluation of water quality is a crucial component of environmental monitoring. Poor water quality damages aquatic life and the environment in its vicinity.	Both studies deal with monitoring water quality. However, the present study is composed of 7 parameters.

Water Parameter Sensors

Water is a necessary element for fish to survive. Due to its wide range of applications, measuring predetermined parameters provides users with real-time online monitoring feedback. Moreover, to monitor the water quality of lakes and rivers and determine the reasons and factors leading to problems with water quality, the device was equipped with a variety of sensors, a communication link, storage and processing capabilities, and energy for powering and utilizing the device (Lee & Lee, 2018).

Dissolved Oxygen (DO) Sensor

The amount of dissolved oxygen (DO) in a body of water is dramatically affected by algae bloom levels. The amount of DO in the water indicates whether the ecosystem has enough oxygen and whether the marine environment is suitable for aquatic organisms' healthy survival. Dissolve Oxygen sensors are commonly used to assess water quality (Menon et al., 2017). Dissolved oxygen sensor measurements are used to track operations where oxygen content influences reaction rates, process efficiency, or environmental conditions. This sensor is commonly utilized when keeping a constant oxygen level, which is critical for maintaining or minimizing reactions (Hong et al., 2021).

pH Level Sensor

One of the most important markers of water quality is pH level. The pH parameters are difficult to measure precisely since the pH level deals with very small amounts of ionic concentration, necessitating a sensitive sensing instrument. A pH sensor is used to offer a general evaluation of water quality. However, the number of sensors can be increased whenever required (Hongpin et al., 2015). The lowest pH value that causes fish death is 4, whereas the maximum is 11. As a result, since it is used to detect the pH value of water, the SEN0161 pH sensor is one of the components in this system. The pH of the water is monitored using these sensors, which are installed at each node (Lauguico et al., 2020).

Temperature Sensor

The water temperature affects fish's appetite and metabolism since it also affects breeding. Moreover, the DS18B20 digital water temperature sensor application is designed and implemented using Python running on the Rpi core controller to send the real-time water temperature read from the sensor. The digital temperature sensor is used to capture the temperature on the Rpi (Zhang et al., 2020). The temperature values of the Dallas Semiconductors DS18B20 are 9 to 12-bit. The sensor communicates using a one-wire interface, connecting several sensors to a single port. The temperature ranges from -55°C to 125°C, with a +/-0.5°C accuracy (Nasution et al., 2020).

Nitrate Sensor

Nitrate (NO3), also referred to as a salt, is a nitrogen and oxygen combination. Nitrogen is released by decomposing matter such as plants, human, and animal waste. Nitrates are necessary for plant growth because they aid in forming amino acids and proteins. Nitrate is extensively water soluble, and excess nitrate that plants do not utilize can seep into groundwater.

The content of nitrate in water is measured with a nitrate meter. Nitrate occurs naturally in water and is not toxic at low concentrations. However, nitrate is dangerous to aquatic ecosystems at high concentrations and, if found in drinking water, can be risky to human health (AquaRead, n.d.).

Phosphorus Sensor

Phosphorus (P) appears in natural water as organic and inorganic phosphates (PO4). Phosphate (PO43) ions are required to generate useful energy from direct sunlight. It promotes cell growth and reproduction. Therefore, a small amount of phosphorus loss from soil may prevent the growth of freshwater weeds and algae. When phosphorus is fed to the lake in excess due to trivial human activities such as urban sewage, street runoff, and rural home septic tanks, it also impacts the environment as well. Phosphorus may travel with groundwater and combine with surface water. As a result, high phosphorus in groundwater may impact surface water quality, which is a major problem. Phosphate levels in water are high, which reduces the amount of dissolved oxygen in lakes and rivers. This phenomenon has a negative impact on the lives of plants and animals within rivers and lakes. Phosphates are not toxic to humans or animals unless present in large quantities. Fish may suffer stomach problems if they live in water with a high level of phosphates. Phosphate levels in fisheries typically range from 0.005 ppm to 0.05 ppm. Periodic blooms are caused by phosphorus levels ranging from 0.08 ppm to 0.10 ppm. However, if phosphate levels continue at 0.05 ppm for an extended period, the eutrophication process will be halted. As a result, for economic and environmental reasons, it is critical to have excellent phosphate management in water used in farming, lakes, or rivers (Akhter et al., 2021). Thus, Phosphorus sensors are designed to monitor the phosphorus level of a given water.

Electrical Conductivity Sensor

An analog electrical conductivity meter tests the electrical conductivity of aqueous solutions and subsequently assesses the water quality. These sensors are commonly used in water culture, aquaculture, environmental detection, and other industries (DFROBOT, n.d.). Additionally, EC sensors are used to test and identify the quality of industrial process water, human drinking water, marine features, and battery electrolytes (Utmel, 2022).

Machine Learning Algorithm

Generally, costly and time-consuming laboratories and statistical analyses have been used to determine water quality. Due to the catastrophic effects of low water quality, a more efficient and affordable approach is required. The study investigates several supervised machine learning algorithm techniques to estimate the water quality class and the water quality index since these are processes that characterize the general quality of water (Harnsoongnoen, 2021).

Random Forest Regression

The most efficient model was created using the Random Forest method, which produced the most trustworthy metrics compared to the other techniques. The percentage errors of predicted and measured DO levels differ significantly when a predictive model is created using the Random Forest Regression (RFR) and Decision Tree Regression (DTR) algorithms (Harrison et al., 2021). Random forest (RF) is a non-linear regression model that can be used in time-series forecasting applications. The RF algorithm consists of an ensemble of decision tree models. The input data is recursively partitioned into two groups based on specified criteria in a decision tree until a predetermined stopping criterion is fulfilled (Insausti et al., 2020).

The study uses a decision tree-based random forest classification method. The idea behind a decision tree algorithm is to train a model by building a series of nodes that divide an observation into different classes based on predictors. The algorithm's objective is to divide the predictors to minimize tree correlation, lowering the procedure's overall variance and improving predictive accuracy. Adding more randomness to the tree-growing process lessens the correlation (Franco, 2019).

RF model to predict algal blooms in the freshwater Urayama Reservoir and the saltwater Lake Shinji, both in Japan. To predict chl-a content one to six months in advance, the researchers analyzed monthly water records with more than ten years' worth of data. The outcomes demonstrated that the RF model could predict the broad trends in chl-a concentration. Additionally, the model enabled the scientists to identify the factors that had the greatest influence on the forecast: pH, total nitrogen/total phosphorus, and chemical and biochemical oxygen demand (Kaidarova et al., 2018).

Methodology

This chapter discussed the research methodology used in the study: research design, procedure, data collection method, sources of data, researcher's instrument construction and validation, distribution, and retrieval of instruments.

Research Design

Developmental Research

The research design that the proponents used in this study is Developmental Research. It is focused on the development and innovation of the device and technology, which is in line with the study's goal. A quantitative method was implemented in the study, as it involved collecting and measuring numerical data and seeking answers to quantitative questions.

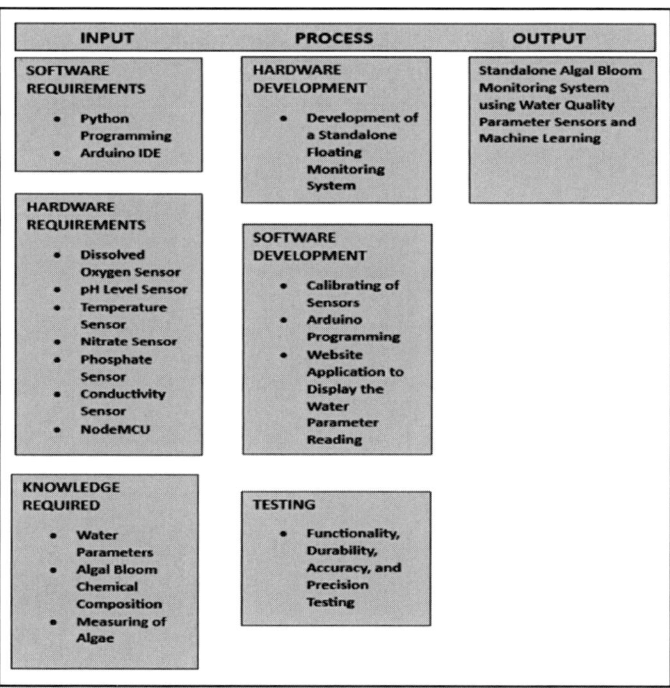

Figure 1. IPO model.

Figures 1 and 2 show the IPO model of the study and its data acquisition, respectively, on which the input contains all the required instruments to be used in building the device. Software requirements are the programs needed to create a web application interface of the device where the data collected by the device would be shown. Similarly, the hardware requirements are for creating Standalone Algal Bloom Monitoring devices. In line with that, water quality parameter sensors are connected to Arduino Uno. Software development is necessary to accurately measure and gather data to predict algal bloom occurrence.

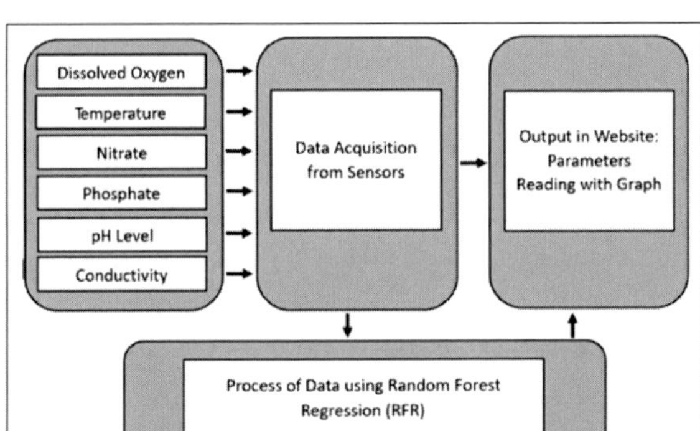

Figure 2. Data acquisition IPO.

Research Flow

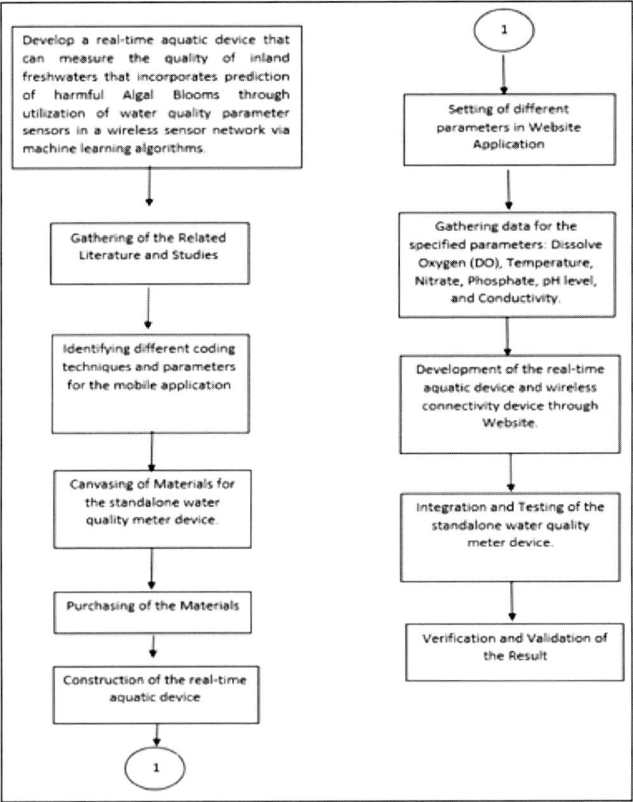

Figure 3. Research flow.

Figure 3 demonstrates how the research process flow actually occurred in sequential sequence throughout the study. The process started with conceptualizing how the real-time monitoring aquatic device came to be the main area of the concentration of the study. The researchers examined studies that had already been completed in order to determine what needed to be altered and modified. Upon constructing the idea for the device, the researchers started setting different parameters in the web application. Lastly, after the testing and integration of the real-time monitoring aquatic device, the validation and verification of the results.

Water Monitoring Buoy with Sensors

This section presented the different methodologies in the development of the hardware prototype, concepts, and parameters for the standalone algal bloom prediction system utilizing water quality sensors. The hardware requirements of the prototype are discussed in this section.

Specifically, the standalone algal bloom monitoring system that was developed is a water monitoring type of buoy since it utilized water quality parameters to predict the harmful algal bloom. A water monitoring buoy, in essence, measures different aspects of water, including water quality, currents, and waves. Among the parts of the aforementioned device are a data recorder, a buoy platform, telemetry gear, solar power, temperature string, mooring hardware, sensors, and sondes. The ideal buoy for a certain application could be made by adjusting the instrumentation and features of a water monitoring buoy station.

Design of the Standalone Algal Bloom Monitoring Buoy

The buoy is designed in consideration of the specifications observed and consolidated with design considerations from the experts. The buoy is quadratically shaped to have stable buoyancy in water. The buoy comes with a considerably heavy load at the bottom of it for stability purposes as well.

To ensure the safety of the sensors that were submerged in water, it had a plastic enclosure with tiny holes on its side. Fish typically attack prey smaller than them and avoid prey significantly larger than them, according to the University of Wisconsin Sea Grant Institute. Hence, as the fish of the beneficiary's pond or a specific body of water are typically Tilapia, Bangus, and other smaller fish, the boy's size is designed to be bigger than those fishes. Figure 4 shows the designed buoy.

Figure 4. Cynosens device.

Dimension of the Buoy

The buoy is composed of a solar panel with the dimension of 9 ⅛ " x 1'1 ⅞ ", and the container which protects the sensors with the dimension of 9 " x 6 " . The dimension of the top view of the buoy is 2 '2" x 2 '2". The side view dimension is 1'5 3/16" x 11".

Research Procedure

Materials and Equipment

Table 4 shows the materials and equipment used such as sensors, microcontrollers, power systems, and peripherals.

Table 4. Materials and equipment

Sensor	Microcontroller	Power System	Peripherals
Dissolved Oxygen Level Sensor	Arduino Uno	Powerbank	Connecting Wires
Temperature Sensor	NodeMCU	Solar Panel	Device Chassis and Insulation
pH level Sensor			
Nitrate and Phosphorus Sensor (NPK)			
Electrical Conductivity Sensor			

Sensors

Dissolved Oxygen Sensor

Dissolved oxygen (DO) sensor can be designed for biochemical oxygen demand (BOD) testing, spot sampling, or long-term monitoring applications. A data logging system, water quality sonde, or dissolved oxygen meter can record measurement information obtained with a DO sensor. A DO sensor measures the amount of oxygen that has been dissolved directly into the water. Dissolved oxygen is crucial for the survival of fish and other aquatic creatures, making it one of the most significant indicators of water quality. Fish, plants, and other aquatic species cannot survive in water that has too little dissolved oxygen. Figure 5 shows the image of the dissolved oxygen sensor that will be used.

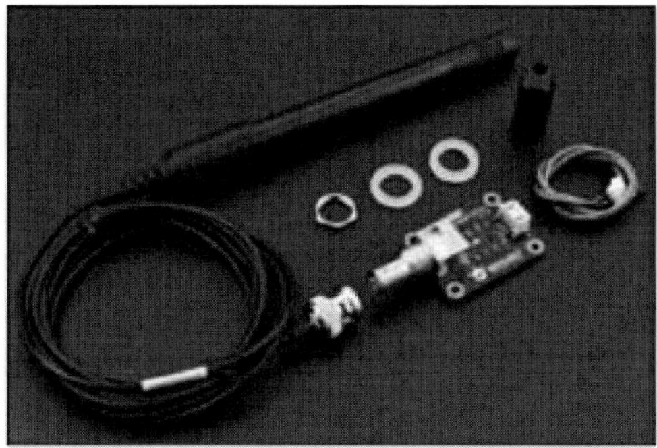

Figure 5. Dissolved oxygen sensor.

pH Level Sensor

A pH Level sensor is one of the most significant devices frequently used to test water. This type of sensor can determine the acidity and alkalinity of water and other fluids. When used effectively, pH sensors can ensure the production processes in a manufacturing or wastewater facility, as well as the safety and quality of a product. The SEN0169 analog pH sensor is designed specifically to determine a solution's acidity or alkalinity by determining its pH. Aquaponics, aquaculture, and environmental water testing are just a few of the uses for this sensor. Figure 6 shows the image of the pH level sensor that was used.

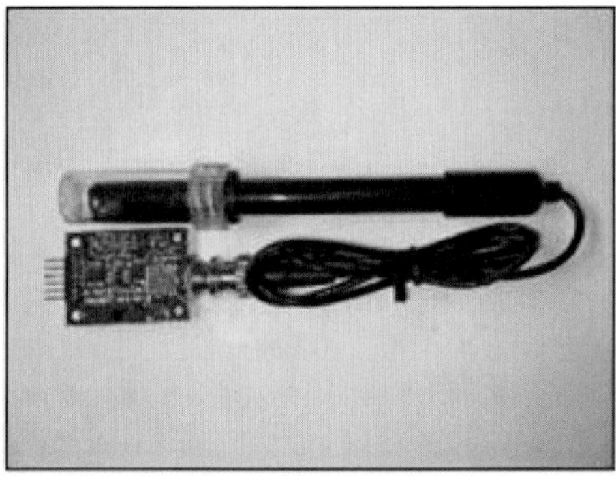

Figure 6. pH level sensor.

Temperature Sensor

A temperature sensor is used to measure the temperature of water. Maxim Integrated makes the DS18B20, a 1-wire programmable temperature sensor. To monitor temperature in adverse circumstances, it is commonly utilized. The DS18B20 sensor can measure a wide range of temperatures, from -55° to +125°, with a reasonable accuracy of 5°C. Because each sensor has a unique address and only takes one MCU pin to relay data, it is a very suitable solution for measuring temperature at numerous locations without using up a lot of your digital pins. The image of the temperature sensor that was employed is depicted in Figure 7.

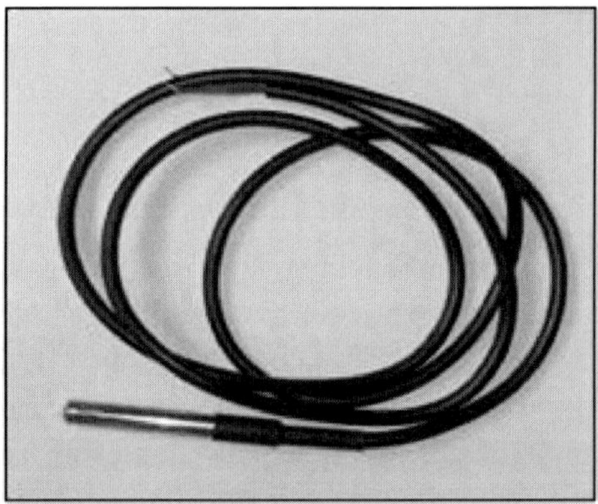

Figure 7. Temperature sensor.

Nitrate and Phosphorus Sensor

NPK sensor as shown in Figure 8 measures the nitrogen, phosphorus, and potassium level and is also used in many different applications such as aquaponics projects and systems. NPK Sensor and Arduino may be used to easily measure the amount of nutrients. To establish the level of nutrient present in soil or water, the content of N (nitrogen), P (phosphorus), and K (potassium) must be measured. This sensor assisted the user in determining nutrient shortfall or abundance.

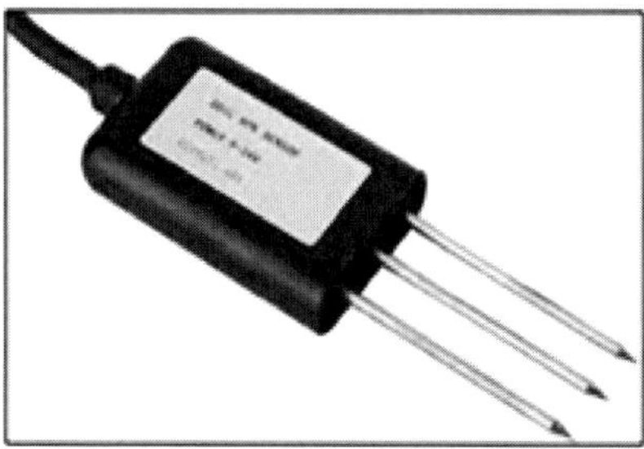

Figure 8. NPK sensor.

Electrical Conductivity Sensor

The DFRobot Electrical Conductivity Sensor as shown in Figure 9, in particular, covers a 3 to 5 volt input range and is compatible with both 3.3 and 5 volt main control boards. The output signal has low jitter and is hardware filtered. An alternating current signal is used as the excitation source, which improves precision, prolongs probe life, and effectively removes the polarization effect. The software library uses a two-point calibration mechanism and has the ability to automatically identify the standard buffer solution, giving the user a straightforward and practical sensor. With this sensor, a primary control board (such an Arduino), and the software library, it is possible to quickly construct an electrical conductivity meter and start plugging it in. In addition to meeting the requirements of various water quality tests, DFRobot provides a variety of water quality sensor devices with consistent sizes and interfaces that are suitable for home-made multi-parameter water quality testers. Conductivity, which relates to a material's capacity to carry electricity, is the inverse of an object's resistivity. The conductivity of a liquid solution is a measure of its capacity to conduct electricity. Conductivity is an important water quality metric. It can reflect the concentration of electrolytes in water (DFROBOT, n.d.).

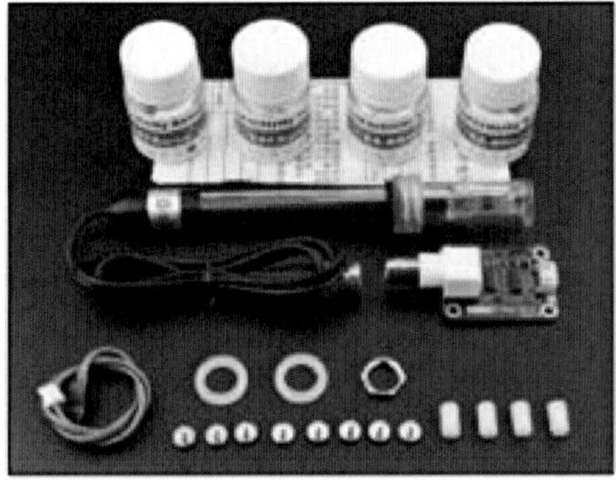

Figure 9. Analog electrical conductivity sensor.

Microcontroller

Arduino UNO

The Arduino UNO as shown in Figure 10 is an open-source programmable microcontroller board that is inexpensive, versatile, and user-friendly and can be used in a wide range of electronic applications. This board can communicate with other Arduino boards, Arduino shields, and Raspberry Pi boards as well as control relays, LEDs, servos, and motors as output devices.

Figure 10. Arduino UNO.

NodeMCU esp8266

The ESP8266, a low-cost System-on-a-Chip (SoC) in Figure 11, is the foundation of the NodeMCU (Node MicroController Unit), an open-source software and hardware development environment. All basic computer parts, including CPU, RAM, networking (WiFi), and even a modern operating system and SDK, are included in the Espressif Systems ESP8266. It is therefore an excellent option for all kinds of Internet of Things (IoT) projects.

Figure 11. NodeMCU esp8266.

Power System

Powerbank

A power bank is a portable charger that enables on-the-go recharging of electrical gadgets. Small, portable devices to larger, higher-capacity Power Banks are available in a variety of sizes. In a similar way, powerbanks are utilized to recharge tablets, GoPros, portable speakers, GPS units, smartphones, cameras, and MP3 players. A Power Bank can be used to charge every device that uses USB power. Only regular charging is required for it as well. A solar panel is then attached to the powerbank so that it may recharge on its own. The image of the powerbank that was used is depicted in Figure 12.

Figure 12. Romoss power bank.

Solar Panel

A solar panel is one aspect of a photovoltaic system that transforms sunlight, composed of energy particles known as "photons," into electricity. They are constructed from a collection of solar cells organized in a panel. They work together to generate electricity and come in a range of rectangular shapes. The power bank's USB port would be used to connect the solar panel. The image of the solar panel that was employed is depicted in Figure 13.

Figure 13. CL-1615 16V 15W solar panel.

Peripherals

Connecting Wires

Jumper wires will be used in connecting circuit boards, components, and other modules to its designated port of connection within the circuit. Figure 14 shows the image of the jumper wires that were used.

Figure 14. Jumper wires.

Device Chassis and Insulation

Plywood in Figure 15 was used as the main material for the chassis of the device. Specifically, it is the main frame of the buoy. Plywood is a lightweight and moisture resistant board. It can be cut, routed, machined, heat-formed or bonded. Plywood features High Strength and Dimensional Stability, Water and Chemical Resistance. Because of its helpful features such as moisture resistance and high strength, plywood has become one of the most preferred building materials. Despite its strength and versatility, plywood is still an economical and long-lasting material for small businesses.

Figure 15. Plywood.

Water-based acrylic-latex paints are often the easiest to work with on plywood projects. Yellow paint was being used to the buoy to easily recognize it when afar in the water.

Figure 16. Paint.

Wall putty is a premium quality white cement-based putty with unique properties and silicone additives and waterproofing properties. Wall putty and epoxy were used to strengthen the chassis of the buoy and make it more water-resistant. Figure 17 shows the image of the Wall Putty and epoxy.

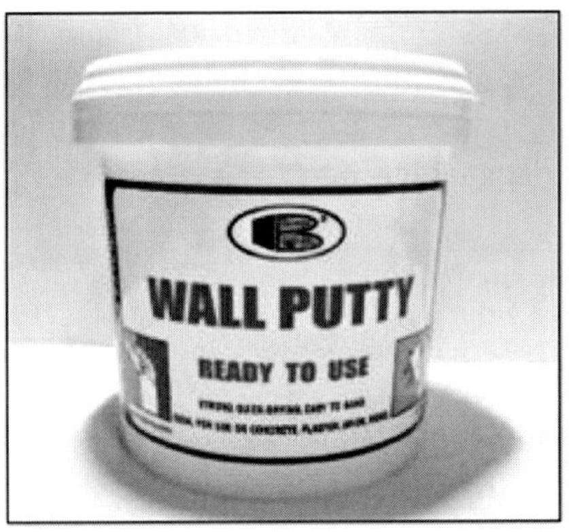

Figure 17. Wall putty (left) and epoxy (right).

Foam was used to cover the inner surface of the chassis to lessen the heat absorbed by the device once it was deployed in a fishpond. Insulation foam is an excellent material used as thermal insulation for roofs, walls, etc. It is made from polyethylene, lightweight and flexible, making it easy to install on roofs and walls. Its closed cell feature makes it water and moisture resistant. Additionally, insulation foam is non fibrous and non-dusting. Also, it is made with fire retardant additives with a self-extinguishing feature. Figure 18 shows the image of the insulation foam that was used.

Figure 18. Insulation foam.

A floating device in Figure 19 is salient to keep the buoy floating in the water. It was intended to give more buoyancy and help the buoy stay afloat in water.

Figure 19. Floating device.

Development of a Program to Predict the Harmful Algal Bloom Using Random Forest Regression

The researchers would predict the occurrence of algal bloom by monitoring the Dissolved Oxygen, Nitrate, pH, Water Temperature, Conductivity, and Phosphate levels present in the water. When it comes to the degree of importance of the parameters, Dissolved Oxygen, Nitrate, and Phosphate are the parameters mostly associated when Algal Bloom is present in water. Nitrate and Phosphate are the two main variables that affect eutrophication which leads to Algal Bloom while Dissolved Oxygen is consumed by bacteria when Algal Bloom is in decomposition state. Random Forest Regression algorithm was utilized because of its versatility, and it is the most reliable method for prediction.

Calibration of Water Quality Parameter Sensors

Various options for sensor calibration are considered to minimize the cost. The researchers first opt to go to different testing laboratories such as BFAR and LLDA that conduct water quality testing to calibrate the sensors that will be used. If there are no available laboratories for testing, purchasing different sensors would be the other option for the basis of sensor calibration.

Data Gathering for Training and Testing Sample

Machine learning models must digest large amounts of structured training data in order to build intelligent applications capable of understanding. The first step in solving any AI-based machine learning problem is to collect enough training data.

Various data sets were gathered and split into the training and testing set in Random Forest Regression.

Development of a Website Application to Display the Acquired Data

Python is a high-level, interpreted, general-purpose programming language used for software development, data analysis and visualization. Python Programming was utilized to develop the web application for the device. The website application displays the different parameter readings. Wireless connection between the device and the web application was established using ESP8266 Wifi Module from Arduino Uno via Internet of Things.

Device Functionality Testing

The researchers tested the device's accuracy by conducting various tests in a controlled environment. The results of the tests were compared to the measurements gathered from existing Algal Bloom monitoring devices or laboratory testing. The device was calibrated until the lowest margin of error was reached for the comparison of the results.

Research Locale

In this study, the researchers envisioned deploying the project in a small-scale aquaculture industry. This project gave a huge help to fisherfolks in monitoring the parameters of their fishponds and alike and knowing ahead of time when the occurrence of harmful algal bloom. With the use of this device, aquaculture farmers can take precautionary actions to protect their livelihood.

The researchers coordinated with the Bureau of Fisheries and Aquatic Resources, Region IV-A in Los Banos, Laguna, to secure a location where the device was safely and effectively deployed for initial stages and training. Additionally, the project has been deployed in a fishpond located at Laguna Lake to further test the accuracy. This collaboration with them was intended to strengthen and solidify the study's conceptual foundations regarding pertinent procedures and relationships critical to the study.

Testing of Device

Accuracy Testing

The accuracy of the proposed device was tested by comparing the gathered data to the readings of EXO YSI Water Monitoring Device of BFAR. The proponents calibrated the proposed device to meet the standards of the device from BFAR.

Testing Procedure

For data acquisition, the researchers determined the relationship of each defined parameter to the level of water quality at which the harmful algal bloom occurred. Then, analyzed the values, specifically the last state of water quality before the algae have occurred. This was done by taking the correlation coefficient of those water parameters.

Based on the obtained results, the researchers found suitable input patterns. Consideration of the optimal combination of inputs was necessary to produce a model that accurately predicted harmful algal blooms.

Evaluation Procedure

According to published literature on hydrological model calibration, validation, and application, several techniques are recommended for model performance evaluation of hydrological time series forecasting.

The random forest regression model was being evaluated using statistical analysis. Water parameter sensors were being tested using the evaluation procedure in Table 5, and the gathered data was compared to the industrial monitoring equipment used by the Bureau of Fisheries and Aquatic Resources in Batangas. The floatation device gave enough buoyancy to the circuit compartment, and the support rope from four sides kept them together.

Table 5. Evaluation procedure

Parameters	EXO YSI Water Monitoring Device of BFAR	Proposed Device
Dissolved Oxygen (mg/L)		
Temperature (°C)		
pH Level		
Nitrate (ppm)		
Phosphate (ppm)		
Conductivity (ms/cm)		

User-Acceptance and Field Testing of Algal Bloom Monitoring System

The testing and evaluation of the device functionality occurred at Batangas Inland Fisheries Technology Outreach Station and Laguna Lake, located southwest of Luzon. The testing consisted of both simulations and actual field testing. The researchers coordinated with the Laguna Lake Development Authority and Bureau of Fisheries and Aquatic Resources Region IV-A (BFAR) in order to acquire authorization for field testing and calibration.

Statistical Analysis

In this study, the following statistical measures were considered by the researchers in this study: Coefficient of Determination (R2), Mean Squared Error (MSE), and Root Mean Squared Error (RMSE). The Coefficient of Determination is a numerical measure which determines how well a model explains and predicts future outcomes. The Mean Squared Error measures how concentrated the data is around the best-fit line. The Root Mean Squared Error is a measure of the difference between the model's predicted and measured values. These measures were used to evaluate and analyze the machine learning model for the prediction of the device.

Results and Discussion

This discussed the results gathered from the testing of the preliminary device.

Project Technical and Description

This study is a standalone monitoring buoy that sends the data gathered from six water quality sensors to website application through a wireless connection, whenever the system senses the fluctuation in the parameters and surpass the normal/standard level of the aforementioned parameters it then sends warning notifications to the website to alert the users about the level of the algae in their water system as a feedback mechanism. The buoy is a solar-powered device that automatically charges the battery embedded in the system.

The prediction system uses Random Forest Regression machine learning algorithms. Random Forest Regression is used in many studies to predict certain parameters as set to it based on the other features associated with the model. The sensors gather the water parameters data continuously and the MySQL database fetched the data from Arduino to save it and display it to the website parameter table for the user's end for monitoring.

In the processing stage, the Random Forest Regression processes the data received by the database to predict the level of the algae in the water and analyze whether it will be harmful to the water system. The algorithm is pre-trained to correlate the features of the six water parameters to the algal bloom level.

For the reflection and feedback stage, the reading of the sensors and warning of the algorithm was displayed and reflected through the website. The website shows the gathered data by the sensors with its chemical unit real time through the cloud base. The notification would appear at the lower right of the website, green colored text was shown if the state is normal otherwise red for bad water quality that could lead to rapid overgrowth of algae in the water. The web application can be used using desktop and mobile view.

Project Structural Organization

Parts of the System

The system is equipped with six water quality sensors namely: Temperature, pH, Nitrate, Phosphorus, Dissolved Oxygen, and Electrical Conductivity. It also has microcontrollers, 20000 mAh powerbank, sensors' module, rechargeable 12 V battery, and Nodemcu esp8266. The six water sensors gather the chemical properties of the water that are mainly involved in the occurrence of harmful algal bloom, it would process to be displayed on the be app.

The laptop unit comes with 512 Gb of SSD, with Windows 11 as an operating system. The computer is equipped with a 6 cores 12 threads 11th Gen Intel(R) Core (TM) i5-1135G7 processor. Running at a base clock speed of 2.42 GHz, 12 Gb of 3733 MHz ram at dual-channel configuration, and an Intel(R) Iris(R) Xe Graphics. The device served as the main processing unit for the Random Forest Regression algorithms.

The microcontroller used by the researchers for the project is Arduino Uno R3. The Arduino Uno is responsible for controlling the sensors' analog and digital inputs and sending them to Nodemcu esp8266.

The solar panel is mounted on the buoy's cover, which converts sunlight into electrical energy to supply power to the battery. The main circuit is placed in the main body of the buoy, also called the circuit compartment, and it is tied to the floating device. The circuit compartment has insulation foam to protect the circuit from heat, and the outside body is covered by epoxy. The circuit compartment box has one hole that is securely perforated in the middle for the sensors to be able to access the water below the buoy. The monitoring process is handled by the buoy, or the hardware device and the web application is responsible for the feedback mechanism and algorithm process of the system. The images below provide a visual representation of the prototype hardware and software project, as shown in Figure 20, CynoSens Monitoring Buoy in Fishpond, and Figure 21, Website View Using Laptop.

Figure 20. CynoSens monitoring buoy in fishpond.

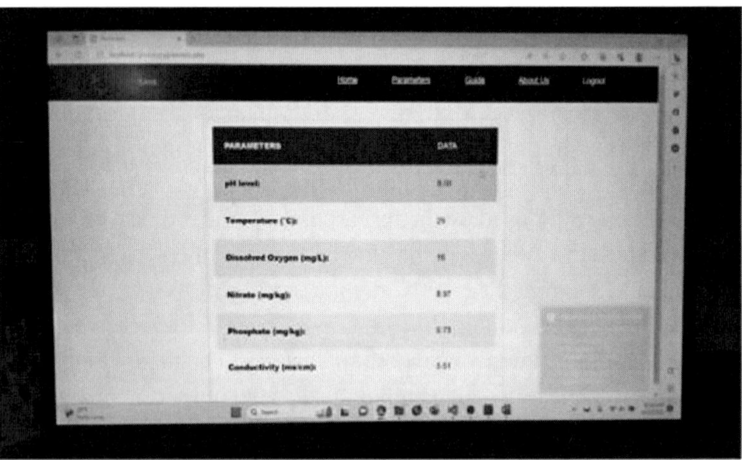

Figure 21. Website view using laptop.

Experimental Result and Data Analysis

Testing – Real Time Gathered Data

Figure 22 shows a real-time sample of data recorded in the database from the Bureau of Fisheries and Aquatic Resources testing site in Tanauan, Batangas. The device was placed in a fishpond and operated during the daytime over two days to monitor water quality parameters and algae content. The figure highlights the gathered data, particularly the low dissolved oxygen levels.

id	phlevel	temperature	dissolvedoxygen	nitrate	phosphate	conductivity
1	7.82	27	8	8.52	0.64	3.48
2	7.45	27	9	8.05	0.96	3.80
3	7.40	29	10	8.26	0.59	3.96
4	7.23	27	10	7.42	0.50	4.36
5	7.29	27	10	8.40	0.54	4.78
6	7.97	27	8	7.42	0.73	4.02
7	7.15	27	10	8.19	0.82	4.99
8	7.56	29	9	8.96	0.80	4.01
9	7.82	29	10	8.04	0.53	3.05
10	7.76	27	9	7.64	0.97	4.56
11	7.19	27	10	8.07	0.85	4.98
12	7.27	29	8	8.71	0.51	4.28
13	7.83	28	10	8.04	0.62	4.57
14	7.65	29	8	8.62	0.86	4.88
15	7.12	29	10	7.74	1.00	4.83
16	7.46	29	8	8.79	0.57	3.77
17	7.50	28	8	8.35	0.88	3.11
18	7.01	29	9	8.59	0.88	4.52
19	7.45	27	10	8.06	0.62	4.37
20	7.69	28	10	7.65	0.95	4.28
21	7.65	28	8	8.30	0.94	3.42
22	7.31	28	8	8.32	0.98	4.49
23	7.27	28	9	7.04	0.75	3.10
24	7.75	27	9	8.85	0.69	4.84
25	7.64	27	9	7.16	0.80	3.49

Figure 22. (Continued).

id	phlevel	temperature	dissolvedoxygen	nitrate	phosphate	conductivity
25	7.64	27	9	7.16	0.80	3.49
26	7.26	28	10	8.33	0.65	4.92
27	7.26	28	9	7.22	0.69	3.55
28	7.75	28	10	7.91	0.71	3.51
29	7.59	29	9	7.14	0.57	3.01
30	7.60	29	9	7.84	0.60	3.86
31	7.38	27	9	7.08	0.67	3.51
32	7.57	27	10	8.91	0.58	3.19
33	7.94	28	9	8.91	0.75	3.56
34	7.11	29	8	7.15	0.88	4.69
35	7.72	28	8	8.96	0.70	4.23
36	7.79	28	10	8.45	0.94	4.82
37	7.24	28	9	7.06	0.56	4.49
38	7.35	27	8	8.70	1.00	4.22
39	7.02	27	10	8.78	0.86	3.87
40	7.74	28	10	8.03	0.75	3.34
41	7.90	29	10	8.76	0.84	3.42
42	7.32	28	8	7.58	0.76	4.68
43	7.99	29	8	7.83	0.89	3.79
44	7.31	28	9	7.98	0.97	3.28
45	7.98	28	8	8.43	0.82	4.24
46	7.45	29	10	8.62	0.67	3.47
47	7.21	27	10	8.65	0.72	3.63
48	7.50	27	9	8.97	0.98	4.39
49	7.00	29	8	8.09	0.65	3.58

Figure 22. Sample of gathered data reflected in database – Batangas.

Deployment – Real Time Gathered Data

Figure 23 shows a real-time sample of data recorded in the database from the deployment site in Laguna Lake. The device was placed in a fishpond and operated for several days to monitor water quality parameters and algae content. As shown in Figure 24, the sample of gathered data is reflected on the website. Figure 25 presents the laboratory test results from the Bureau of Fisheries and Aquatic Resources (BFAR), confirming that the algae content remained at a normal level of approximately 15 μg/L. The data collected from sensors is presented in Figure 26, which features the EXO YSI Water Monitoring Device used by BFAR.

id	phlevel	temperature	dissolvedoxygen	nitrate	phosphate	conductivity	algaecontent	status	time
26901	7.61	26	6	7.39	0.57	0.69	15.34	Algae Level: Normal	2023-07-05 01:40:35
26902	7.44	26	6	7.32	0.56	0.66	15.30	Algae Level: Normal	2023-07-05 01:42:02
26903	7.51	26	6	7.43	0.54	0.66	15.75	Algae Level: Normal	2023-07-05 01:42:32
26904	7.46	26	6	7.32	0.59	0.69	15.89	Algae Level: Normal	2023-07-05 01:43:02
26905	7.69	26	6	7.59	0.55	0.70	15.49	Algae Level: Normal	2023-07-05 01:43:32
26906	7.64	26	6	7.39	0.56	0.68	15.75	Algae Level: Normal	2023-07-05 01:44:02
26907	7.40	26	6	7.34	0.54	0.70	15.42	Algae Level: Normal	2023-07-05 01:45:02
26908	7.21	26	7	7.39	0.58	0.66	15.55	Algae Level: Normal	2023-07-05 01:45:32
26909	7.40	26	7	7.47	0.57	0.67	15.41	Algae Level: Normal	2023-07-05 01:46:02
26910	7.44	26	7	7.34	0.54	0.70	15.79	Algae Level: Normal	2023-07-05 01:46:32
26911	7.42	26	7	7.47	0.54	0.70	15.32	Algae Level: Normal	2023-07-05 01:47:02
26912	7.64	26	6	7.56	0.59	0.69	15.61	Algae Level: Normal	2023-07-05 01:47:32
26913	7.60	26	7	7.31	0.58	0.70	15.76	Algae Level: Normal	2023-07-05 01:48:03
26914	7.64	26	7	7.56	0.54	0.69	15.84	Algae Level: Normal	2023-07-05 01:48:33
26915	7.45	26	7	7.58	0.57	0.71	15.34	Algae Level: Normal	2023-07-05 01:49:03
26916	7.52	26	6	7.44	0.55	0.70	15.40	Algae Level: Normal	2023-07-05 01:49:33
26917	7.21	26	7	7.35	0.59	0.66	15.89	Algae Level: Normal	2023-07-05 01:50:03
26918	7.60	26	7	7.30	0.59	0.69	15.70	Algae Level: Normal	2023-07-05 01:50:33
26919	7.43	26	7	7.57	0.54	0.69	15.54	Algae Level: Normal	2023-07-05 01:51:33
26920	7.20	26	7	7.30	0.56	0.68	15.30	Algae Level: Normal	2023-07-05 01:52:03
26921	7.39	26	7	7.32	0.57	0.69	15.80	Algae Level: Normal	2023-07-05 01:52:33
26922	7.53	26	6	7.49	0.54	0.68	15.76	Algae Level: Normal	2023-07-05 01:53:03
26923	7.70	26	7	7.36	0.59	0.70	15.81	Algae Level: Normal	2023-07-05 01:53:33
26924	7.42	26	7	7.34	0.56	0.66	15.40	Algae Level: Normal	2023-07-05 01:54:03
26925	7.48	26	6	7.49	0.54	0.70	15.40	Algae Level: Normal	2023-07-05 01:54:33
26926	7.42	26	6	7.51	0.55	0.70	15.37	Algae Level: Normal	2023-07-05 01:55:03
26927	7.30	26	7	7.34	0.55	0.70	15.74	Algae Level: Normal	2023-07-05 01:55:33
26928	7.21	26	7	7.40	0.56	0.66	15.74	Algae Level: Normal	2023-07-05 01:56:03
26929	7.60	26	6	7.41	0.54	0.66	15.61	Algae Level: Normal	2023-07-05 01:57:03
26930	7.29	26	7	7.50	0.55	0.67	15.50	Algae Level: Normal	2023-07-05 01:57:33

Figure 23. Sample of gathered data reflected in database – Laguna Lake.

PARAMETERS	DATA
ID Number:	27222
pH Level:	7.51
Temperature (°C):	26
Dissolved Oxygen (mg/L):	6
Nitrate (ppm):	7.34
Phosphate (ppm):	0.57
Conductivity (ms/cm):	0.68
Algae Content (ug/L):	15.11
Status:	Algae Level: Normal
Date and Time:	2023-07-05 04:44:33

Figure 24. Sample of gathered data reflected on website.

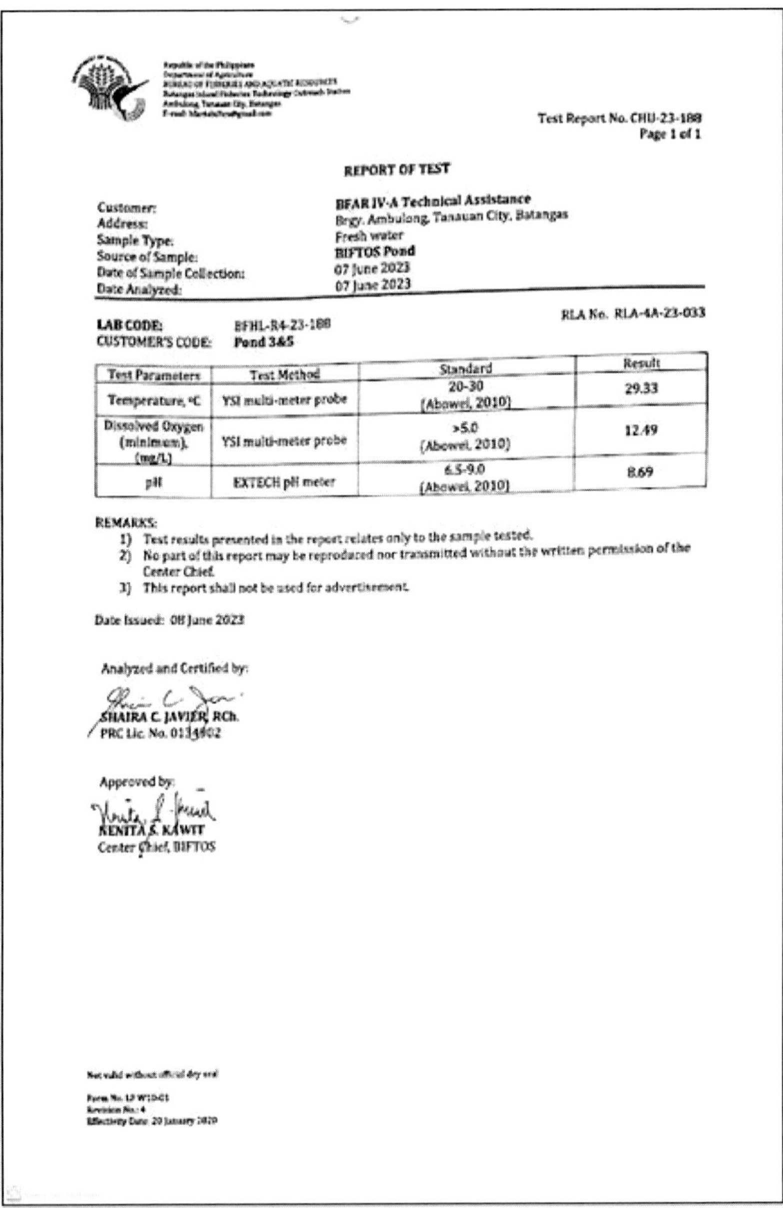

Republic of the Philippines
Department of Agriculture
BUREAU OF FISHERIES AND AQUATIC RESOURCES
Batangas Inland Fisheries Technology Outreach Station
Ambulong, Tanauan City, Batangas

Test Report No. CHU-23-188
Page 1 of 1

REPORT OF TEST

Customer: **BFAR IV-A Technical Assistance**
Address: Brgy. Ambulong, Tanauan City, Batangas
Sample Type: Fresh water
Source of Sample: **BIFTOS Pond**
Date of Sample Collection: 07 June 2023
Date Analyzed: 07 June 2023

RLA No. RLA-4A-23-033

LAB CODE: BFHL-R4-23-188
CUSTOMER'S CODE: **Pond 3&5**

Test Parameters	Test Method	Standard	Result
Temperature, °C	YSI multi-meter probe	20-30 (Abowei, 2010)	29.33
Dissolved Oxygen (minimum), (mg/L)	YSI multi-meter probe	>5.0 (Abowei, 2010)	12.49
pH	EXTECH pH meter	6.5-9.0 (Abowei, 2010)	8.69

REMARKS:

1) Test results presented in the report relates only to the sample tested.
2) No part of this report may be reproduced nor transmitted without the written permission of the Center Chief.
3) This report shall not be used for advertisement.

Date Issued: 08 June 2023

Analyzed and Certified by:

SHAIRA C. JAVIER, RCh.
PRC Lic. No. 0134402

Approved by:

NENITA S. KAWIT
Center Chief, BIFTOS

Not valid without official dry seal

Form No. [illegible]
Revision No.: 4
Effectivity Date: 20 January 2020

Figure 25. Laboratory test from BFAR.

Figure 26. EXO YSI water monitoring device of BFAR.

Table 6 shows the gathered data by the device and the data resulted from the expert analysis. The device is tested at the Bureau of Fisheries and Aquatic Resources in Tanauan, Batangas. The monitoring device and the prediction buoy are placed in the water simultaneously to collect the water parameters. During the gathering of data, the researchers noted that they gathered data from their standard monitoring device to the prediction buoy system.

Table 6. BFAR gathered data by the EXO YSI device vs. CynoSens

BFAR	Cynosens
Temperature: 29.33°C	Temperature: 30°C
Dissolved Oxygen: 12.49 mg/L	Dissolved Oxygen: 11.89 mg/L
pH: 8.69	pH: 8.5
Nitrate: 8.62 ppm	Nitrate: 8.97 ppm
Phosphate: 0.86 ppm	Phosphate: 0.70 ppm
	Conductivity: 4.88 us/cm

Evaluation of Random Forest Regression Prediction Model

In this study, Random Forest Regression machine learning algorithm was used to predict the alga content of the water using the other features which in this case are the water parameters. The created model is trained first using the dataset acquired and then hyper tuned using randomized search CV to enhance the predicted values it outputs. After hypertuning, the model was tested using the gathered values and evaluates its level of prediction.

The dataset has been split into training data (80%) and test data (20%). In the random forest regressor, the decision tree scales the input. The model is demonstrated by fetching the data of feature water parameters by the sensors in the database and predicting the Algae content using an HTML file. The file is put in a pickle file and the performances of the model are computed.

Figure 27 shows the sample result of prediction during the testing of the algorithm. The algorithm gathered the features then predicted the algae content value and compared it to the target or actual value.

```
In [25]: df_eval = pd.DataFrame(rf_random.predict(X_test), columns=['Prediction'])
         y_test = y_test.reset_index(drop=True)
         df_eval['Target'] = y_test

         df_eval['Residual'] = df_eval['Target']-df_eval['Prediction']
         df_eval['Differences%'] = np.absolute(df_eval['Residual']/df_eval['Target']*100)
         df_eval
Out[25]:
```

	Prediction	Target	Residual	Differences%
0	9.5290	4.1	-5.4290	132.414634
1	1.2464	1.1	-0.1464	13.309091
2	5.6760	6.7	1.0240	15.283582
3	18.2490	23.9	5.6510	23.644351
4	0.5892	0.0	-0.5892	inf
...				
125	3.7120	7.2	3.4880	48.444444
126	16.2820	10.5	-5.7820	55.066667
127	6.7000	8.0	1.3000	16.250000
128	11.8134	23.7	11.8866	50.154430
129	4.3710	1.5	-2.8710	191.400000

130 rows × 4 columns

Figure 27. Sample result of the Testing Prediction of the Random Forest Regression Algorithm.

The performance measures are computed using the R squared Score, Root mean square error and Mean Square Error. The model's accuracy is determined using the built-in functions of Sklearn. If there is a larger than cutoff (MSE) gap between the test and the predicted value, the prediction was incorrect, if not, accurate prediction.

The accuracy of the model calculated resulted in a score of 0.9148247255739806, hence the algorithm model accuracy is 91.48%.

Website Application Interface

The web application as shown in Figure 28 displays the value of each parameter from the hardware sensors. The website can be accessed through any device because of its responsiveness which makes the display of the website fit the size of the screen of the device. This allows the website to be easily accessible to the users. The website contains: login and register, homepage, parameters, guide, and about us.

Figure 28. Login page (computer view).

PARAMETERS	DATA
ID Number:	11540
pH Level:	7.92
Temperature (°C):	27
Dissolved Oxygen (mg/L):	8
Nitrate (ppm):	7.87
Phosphate (ppm):	0.59
Conductivity (ms/cm):	0.67
Algae Content (ug/L):	16.59
Status:	Algae Level: Normal
Date and Time:	2023-06-28 19:22:02

Figure 29. Parameters (computer view).

Figures 29 and 30 showcase a designed parameters webpage that offers comprehensive insights. Within this webpage, a table is thoughtfully presented, featuring the six (6) water quality parameters alongside their corresponding IDs, dates, and times. Additionally, the table includes crucial information on algae level and content. One of the exceptional features of this webpage is its real-time data display, providing continuous updates every thirty (30) seconds. To ensure users are promptly informed, a synchronized notification pops up simultaneously with each data update. These notifications effectively communicate the parameter levels in terms of low, normal, and high categories, aligning with the standard water quality for type c pond waters from BFAR and the anticipated algae content.

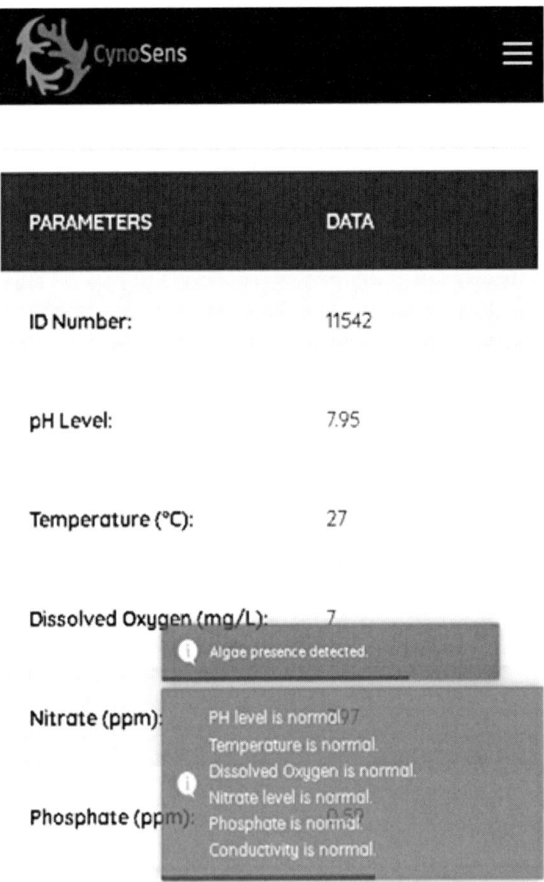

Figure 30. Parameters (smartphone view).

Furthermore, an additional synchronized notification appears to present a prediction derived from an algorithm. This prediction serves to indicate whether an algal bloom is occurring, supplying valuable information regarding the algae level and content, along with an assessment of the water's safety for the pond's fish. The combination of real-time data updates, informative notifications, and predictive capabilities.

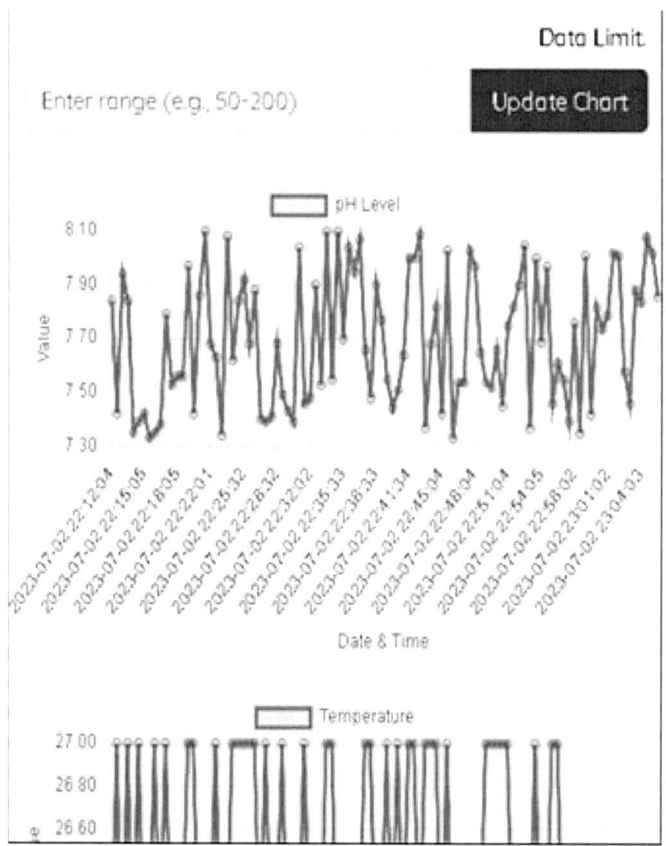

Figure 31. Chart (smartphone view).

As shown in Figure 31, an algae content chart was integrated to visually depict trends and fluctuations in the algae level within the water. Notably, an interactive feature was implemented allowing users to assess the algae level by modifying the color of pointers, based on the legend displayed below the algae content chart. The legend provided color classifications such as blue for low algae levels, green for normal levels, yellow for above-normal levels, and red to indicate the presence of harmful levels of algae. It is worth mentioning that the data was presented in real-time, ensuring accuracy by automatically updating every minute. To facilitate effortless navigation, users could effortlessly hover their mouse cursor on a computer or use their fingers on smartphones to explore specific data points and view the corresponding values. Moreover, users were granted the flexibility to personalize their chart display by selecting a desired data range, choosing which data to be displayed, including or excluding specific information, and even zooming in or out on their smartphones. The amalgamation of real-time updates, interactive features, and user-friendly customization options rendered this chart webpage an exceptional tool for monitoring and analyzing water quality data.

Summary, Conclusion, and Recommendation

This contains the summary of the study, its findings, conclusions drawn, and presents recommendations for utilization and further research and investigation.

Summary

Harmful algal blooms (HABs) seriously threaten aquatic ecosystems and human health. Effective management and mitigation techniques depend on early HAB prediction and monitoring. A network of interconnected sensors placed in bodies of water such as lakes and fishponds make up the IoT-based Harmful Algal Bloom Prediction System. These sensors gathered information about the lake's temperature, pH, dissolved oxygen, conductivity, nitrate, and phosphate. The sensors utilized microcontrollers with wireless communication protocols to transmit the data they gathered to a web database.

The proposed device utilized Random Forest Regression as the machine learning algorithm that combines multiple decision trees to make predictions. It is capable of handling complex and non-linear relationships between input variables and the target variable. In the context of harmful algal bloom prediction, RFR utilizes historical water quality data and associated algal bloom occurrences to train a model that can forecast future bloom events. The gathered data is then transmitted to the database, which is then displayed to the web application where users can view the readings.

The proposed device shows better functionality compared to the other water quality monitoring systems available in the market. Other devices such as the Sonde YSI Monitoring Device monitor water quality but lack wireless capability compared to the proposed project that transmits the data gathered wirelessly to the smartphones or laptops of the fishpond owners. Moreover, nutrient contents of the water such as nitrate and phosphorus are usually tested once a month. With the proposed device, nutrient contents of the water are monitored from time to time with the inclusion of algal bloom prediction through water parameters which is not yet available from currently available products in the market.

Conclusion

The study presented efficient data in monitoring and reading water quality parameters through the use of sensors that aids in the prediction of occurring algal blooms. Based on the results and findings gathered on the research, the following conclusions were made:

1. The researchers established a standalone algal bloom monitoring buoy that utilizes water quality sensors namely: Dissolved Oxygen (DO) sensor, temperature sensor, pH level sensor, conductivity sensor, nitrate sensor, and phosphate sensor. The data collected by the standalone algal bloom monitoring buoy is comparable to laboratory testing, which is the existing water quality monitoring method.
2. The development of a program that predicts the harmful algal bloom is established by using the machine learning algorithm Random Forest Regression. Its real-time data collection enables early detection and response to potential bloom events. The modeling approach of Random Forest Regression ensures reliable and adaptable predictions for various water bodies and environmental conditions. The calculated accuracy of the model resulted in a score of 0.9148247255739806, thus being 91.48% accurate.
3. The website application was also successfully developed using HTML for the front end, CSS and Javascript for the website layout and design. As for the backend, the proponents utilized the PHP server programming language to connect the website to the MySQL database. These programming languages are utilized to display the acquired data from the buoy. The website serves as a centralized

platform that offers valuable information, data visualization, and user-friendly tools to enhance the understanding and management of algal blooms in water bodies. And with its notification, users will always be updated on any changes to the status of the environment.

4. The functionality of the device was proven to operate effectively and demonstrated significant durability through simulations and actual field testing in a fishpond in BFAR Batangas and Laguna Lake. The system was able to accurately measure the level of water quality parameters and detect the level of algae content present in the water greater than 50 μg/L. The algorithm's prediction was triggered at 21 μg/L with an accuracy of 91.48%.

Recommendations

Based on the findings and conclusions of this project, the following recommendations are provided:

1. Develop a cost-efficient chlorophyll-a sensor as it is one of the important sensors that can accurately detect harmful algal bloom.
2. Utilizing a mobile app instead of a webpage. Mobile apps offer offline access and the ability to store data locally on the device, enabling users to access and interact with the app even without an internet connection.
3. To further augment the device by developing a system that can automatically correct the water parameters and is not dependent on human intervention, enabling it to predict algal blooms and actively mitigate their occurrence.

References

Administrator. (2016, November 20). *Arduino Introduction*. ElectronicsHub. https://www.electronicshub.org/arduino-introduction/.

Agarwal, T. (2019, August). *Arduino Mega 2560 Board: Specifications, and Pin Configuration*. ElProCus - Electronic Projects for Engineering Students. https://www.elprocus.com/arduino-mega-2560-board/.

Ahmed, U., Mumtaz, R., Anwar, H., Shah, A. A., Irfan, R., & José García-Nieto. (2019). *Efficient Water Quality Prediction Using Supervised Machine Learning*. *11*(11), 2210–2210. https://doi.org/10.3390/w11112210.

Akhter, F., Hasin Reza Siddiquei, Alahi, E. E., & Subhas Chandra Mukhopadhyay. (2021). *Recent Advancement of the Sensors for Monitoring the Water Quality Parameters in Smart Fisheries Farming*. *10*(3), 26–26. https://doi.org/10.3390/computers10030026.

Altynay Kaidarova, Marengo, M., Marinaro, G., Geraldi, N. R., Duarte, C. M., & Kosel, J. (2018). *Flexible and Biofouling Independent Salinity Sensor*. *5*(23), 1801110–1801110. https://doi.org/10.1002/admi.201801110.

Anderson, D. M. (2009). *Approaches to monitoring, control and management of harmful algal blooms (HABs)*. *52*(7), 342–347. https://doi.org/10.1016/j.ocecoaman.2009.04.006.

Anderson, J. M., Asche, F., & Garlock, T. (2019). *Economics of Aquaculture Policy and Regulation*. Annual Reviews. https://www.annualreviews.org/doi/10.1146/annurev-resource-100518-093750.

Ayele, H., & Minaleshewa Atlabachew. (2021). *Review of characterization, factors, impacts, and solutions of Lake eutrophication: lesson for lake Tana, Ethiopia*. *28*(12), 14233–14252. https://doi.org/10.1007/s11356-020-12081-4.

B, R. (2018, December 14). *What is MySQL: MySQL Explained For Beginners*. Hostinger Tutorials. https://www.hostinger.com/tutorials/what-is-mysql.

Beyeler, R. (2013). *Pocket Wi-Fi*. Moneyland.ch; moneyland.ch. https://www.moneyland.ch/en/pocket-wifi-definition#:~:text=A%20pocket%20Wi%2DFi%20is,can%20connect%20to%20the%20Internet.

Borlaza, G. C., Hernandez, C. G., & Cullinane, M. (2023, July 7) *Aquaculture Pollution | Other Environmental Study*. (2018, February 16). Other Environmental Study. https://elibrary.worldbank.org/doi/abs/10.1596/29249.

Brignell, J. E., & Dorey, A. P. (1983). *Sensors for microprocessor-based applications*. https://doi.org/10.1088/0022-3735/16/10/003.

Demetillo, A. T., Japitana, M. V., & Taboada, E. B. (2019). *A system for monitoring water quality in a large aquatic area using wireless sensor network technology*. *29*(1). https://doi.org/10.1186/s42834-019-0009-4.

Espaldon, M. V. & et al. (2005). Ecosystems and People: The Philippine Millennium Ecosystem Assessment (MA) Sub-Global Assessment. *Sub-Global*, 55–196.

Fabiane Fantinelli Franco, Libu Manjakkal, Dhayalan Shakthivel, & Ravinder Dahiya. (2019). *ZnO based Screen Printed Aqueous Ammonia Sensor for Water Quality Monitoring*. https://doi.org/10.1109/sensors43011.2019.8956763.

Gravity__Analog_Electrical_Conductivity_Sensor___Meter_V2__K=1__SKU_DFR0300-DFRobot. (2018). Dfrobot.com. https://wiki.dfrobot.com/Gravity__Analog_Electrical_Conductivity_Sensor___Meter_V2__K=1__SKU_DFR0300.

Gravity__Analog_Electrical_Conductivity_Sensor___Meter-DFRobot. (2023). Dfrobot.com. https://wiki.dfrobot.com/Gravity__Analog_Electrical_Conductivity_Sensor___Meter.

Gu, J., Börklü, H. R., Helvacilar, E., & Özdemir, V. (2017). *Conceptual Design of a New Buoy* (Vol. 4, Issue 4).

Harrison, J., Lucius, M. A., Farrell, J. L., Eichler, L. W., & Relyea, R. A. (2021). *Prediction of stream nitrogen and phosphorus concentrations from high-frequency sensors using Random Forests Regression*. *763*, 143005–143005. https://doi.org/10.1016/j.scitotenv.2020.143005.

Harrison, J., Lucius, M. A., Farrell, J. L., Eichler, L. W., & Relyea, R. A. (2021). *Prediction of stream nitrogen and phosphorus concentrations from high-frequency sensors using Random Forests Regression. 763,* 143005–143005. https://doi.org/10.1016/j.scitotenv.2020.143005.

Heisler, J., Glibert, P. M., Burkholder, J. M., Anderson, D. Z., Cochlan, W. P., Dennison, W. C., Dortch, Q., Gobler, C. J., Heil, C. A., Humphries, E., Lewitus, A. J., Magnien, R. E., Marshall, H. L., Sellner, K. G., Stockwell, D. A., Stoecker, D. K., & Suddleson, M. (2008). *Eutrophication and harmful algal blooms: A scientific consensus. 8*(1), 3–13. https://doi.org/10.1016/j.hal.2008.08.006.

Kate, S., Santos, B. S., & Magbanua, F. S. (2020). Geometric Morphometric Analysis of Channa striata (Striped Snakehead) Populations from Laguna de Bay, Philippines Reveals Shape Differences in Relation to Water Quality. *Science Diliman: A Journal of Pure and Applied Sciences, 32*(2). https://www.journals.upd.edu.ph/index.php/sciencediliman/article/view/7384.

Ki Wook Kim, Min Whan Jung, Yiu Fai Tsang, & Kwon, H.-H. (2020). *Stochastic modeling of chlorophyll-a for probabilistic assessment and monitoring of algae blooms in the Lower Nakdong River, South Korea. 400,* 123066–123066. https://doi.org/10.1016/j.jhazmat.2020.123066.

Lauguico, S., Concepcion, R., Rogelio Ruzcko Tobias, Alejandrino, J., Macasaet, D., & Dadios, E. P. (2020). *Indirect Measurement of Dissolved Oxygen Based on Algae Growth Factors Using Machine Learning Models.* https://doi.org/10.1109/r10-htc49770.2020.9357014.

Lee, S.-M., & Dong Hyun Lee. (2018). *Improved Prediction of Harmful Algal Blooms in Four Major South Korea's Rivers Using Deep Learning Models. 15*(7), 1322–1322. https://doi.org/10.3390/ijerph15071322.

Lorenzo, A. R., Dula, A. Y., Aldrin, N., John Randolph Munda, Noli, B., Padilla, M., Madrigal, A. M., Amado, T. M., & Karlo, L. (2019). *Dissolved Oxygen (DO) Meter Hydrological Modelling Using Predictive Algorithms.* https://doi.org/10.1109/hnicem48295.2019.9072867.

Luo Hongpin, Guanglin, L., Pan Weifeng, Jie, S., & Bai Qiuwei. (2015). *Real-time remote monitoring system for aquaculture water quality. 8*(6), 136–143. https://doi.org/10.25165/ijabe.v8i6.1486.

Lutkevich, B. (2020). *HTML (Hypertext Markup Language).* TheServerSide.com; TheServerSide.com. https://www.theserverside.com/definition/HTML-Hypertext-Markup-Language.

Mapa, D. (n.d.). *PHILIPPINE STATISTICS AUTHORITY.* Retrieved July 11, 2023, from https://psa.gov.ph/sites/default/files/Fisheries%20Situation%20Report%20for%20Major%20Species%2C%20January%20to%20December%202021.pdf.

Maria Shamina D'Silva, Arga Chandrashekar Anil, Naik, R. K., & D'Costa, P. M. (2012). *Algal blooms: a perspective from the coasts of India. 63*(2), 1225–1253. https://doi.org/10.1007/s11069-012-0190-9.

Matías Insausti, Timmis, R., Kinnersley, R., & Rufino, M. C. (2020). *Advances in sensing ammonia from agricultural sources. 706,* 135124–135124. https://doi.org/10.1016/j.scitotenv.2019.135124.

Mell, P. M., & Grance, T. (2011, September 28). *The NIST Definition of Cloud Computing*. NIST. https://www.nist.gov/publications/nist-definition-cloud-computing.

Mendoza, M. U., et al. (2019). Water quality and weather trends preceding fish kill occurrences in Lake Taal (Luzon Is., Philippines) and recommendations on its long-term monitoring. *Vol. 12, No. 02.*

Menon, G., Maneesha Vinodini Ramesh, & P. Divya. (2017). *A low-cost wireless sensor network for water quality monitoring in natural water bodies*. https://doi.org/10.1109/ghtc.2017.8239341.

Nadarajah, S., & Flaaten, O. (2017). *Global aquaculture growth and institutional quality*. *84*, 142–151. https://doi.org/10.1016/j.marpol.2017.07.018.

Nag, A., Subhas Chandra Mukhopadhyay, & Kosel, J. (2017). *Sensing system for salinity testing using laser-induced graphene sensors*. *264*, 107–116. https://doi.org/10.1016/j.sna.2017.08.008.

Ngatia, L. W., Grace, J. M., Moriasi, D. N., & Taylor, R. W. (2019). *Nitrogen and Phosphorus Eutrophication in Marine Ecosystems*. https://doi.org/10.5772/intechopen.81869.

Nitrate Water Testing Sensors - AquaRead. (2020). Aquaread.com. https://www.aquaread.com/sensors/nitrate.

O'Grady, B. (2020, February 25). *What is CSS? The Ultimate Intro - Code Institute Global*. Code Institute Global. https://codeinstitute.net/global/blog/what-is-css-and-why-should-i-learn-it/.

Prananingtyas, D., Prayogo, & Susanto Rahardja. (2019). *Effect of Different Salinity Level within Water Against Growth Rate, Survival Rate (FCR) of Catfish (Clarias sp.)*. https://doi.org/10.1088/1755-1315/236/1/012035.

Rafaela Costa Cruz, Costa, P., Vinga, S., Ludwig Krippahl, & Julia, M. (2021). *A Review of Recent Machine Learning Advances for Forecasting Harmful Algal Blooms and Shellfish Contamination*. *9*(3), 283–283. https://doi.org/10.3390/jmse9030283.

Rakibul Islam Chowdhury, Wahid, K. A., Nugent, K., & Baulch, H. M. (2020). *Design and Development of Low-Cost, Portable, and Smart Chlorophyll-A Sensor*. *20*(13), 7362–7371. https://doi.org/10.1109/jsen.2020.2978758.

RMBEL (2018, February 1). *Lake Eutrophication - RMBEL*. (2018, February). RMBEL. https://www.rmbel.info/primer/lake-eutrophication/.

Robocraze. (2022, August 4). *What is NodeMCU ESP8266: Complete Guide*. Robocraze; Robocraze. https://robocraze.com/blogs/post/what-is-nodemcu-esp8266.

Supakorn Harnsoongnoen. (2021). *Metamaterial-Inspired Microwave Sensor for Detecting the Concentration of Mixed Phosphate and Nitrate in Water*. *70*, 1–6. https://doi.org/10.1109/tim.2021.3086901.

Tayaban, M. M., Pintor, K. L., & Vital, P. G. (2017). *Detection of potential harmful algal bloom-causing microalgae from freshwater prawn farms in Central Luzon, Philippines, for bloom monitoring and prediction*. *20*(3), 1311–1328. https://doi.org/10.1007/s10668-017-9942-8Techopedia. (2020, August 7). *What is a Mobile Application? - Definition from Techopedia.*

TechTarget Contributor. (2023). *web application (web app)*. Software Quality; TechTarget. https://www.techtarget.com/searchsoftwarequality/definition/Web-application-Web-app.

The Effects: Dead Zones and Harmful Algal Blooms | US EPA. (2013, March 12). US EPA. https://www.epa.gov/nutrientpollution/effects-dead-zones-and-harmful-algal-blooms.

Tigor Hamonangan Nasution, Dika, S., Sinulingga, E. P., Kasmir Tanjung, & Lukman Adlin Harahap. (2020). *Analysis of the use of SEN0161 pH sensor for water in goldfish ponds*. *851*(1), 012053–012053. https://doi.org/10.1088/1757-899x/851/1/012053.

Tiongson, K. N., & Tamayo-Zafaralla, M. (2018). *View of Water Quality and Seasonal Dynamics of Phytoplankton and Zooplankton in the West Bay of Laguna de Bay, Philippines*. (2023). Uplb.edu.ph. https://ovcre.uplb.edu.ph/journals-uplb/index.php/EDJ/article/view/238/222.

Toal, R. (2014, June 10). *What is PHP? Uses & Introduction - Code Institute Global*. Code Institute Global. https://codeinstitute.net/global/blog/what-is-php-programming/.

Truong, P. (2019, February 24). *Unraveling the Web of IoT: The Internet of Things - Phu Truong - Medium*. Medium; Medium. https://medium.com/@truong21/unraveling-the-web-of-iot-the-internet-of-things-669c230fa4aa.

UTMEL. (2022, June 20). *What is Conductivity Sensor?* Utmel.com; Utmel Electronics. https://www.utmel.com/blog/categories/sensors/what-is-conductivity-sensor.

What is the Internet of Things (IoT)? (2020). Oracle.com. https://www.oracle.com/ph/internet-of-things/what-is-iot/.

Wilhelm, F. M. (2009). *Pollution of Aquatic Ecosystems I*. 110–119. https://doi.org/10.1016/b978-012370626-3.00222-2.

Wong Jun Hong, Shamsuddin, N., Pg Emeroylariffion Abas, Rosyzie Anna Apong, Masri, Z., Hazwani Suhaimi, Gödeke, S., & Muhammad. (2021). *Water Quality Monitoring with Arduino Based Sensors*. *8*(1), 6–6. https://doi.org/10.3390/environments8010006.

Wurtsbaugh, W. A., Paerl, H. W., & Dodds, W. K. (2019). *Nutrients, eutrophication and harmful algal blooms along the freshwater to marine continuum*. *6*(5). https://doi.org/10.1002/wat2.1373.

Zach Paruch. (2021). Semrush Blog. https://www.semrush.com/blog/javascript/.

Zhang, Z., Mao, W., Wang, Z., Tan, X., Wu, F., Wang, D., & Fang, X. (2020). *Development of remote monitoring system for aquaculture water quality based on Internet of Things*. https://doi.org/10.1088/1757-899x/768/5/052033.

Zola, A. (2021). *Python*. WhatIs.com; TechTarget. https://www.techtarget.com/whatis/definition/Python.

Chapter 4

SMAQ: A Smart Aquaponics System Using IoT Technology

Edmon O. Fernandez[1,2,*]
Jomer V. Catipon[1,2]
Reign Nicole G. Mabalay[1]
Wyen Cazmyr C. Cabrito[1]
Rommel A. Jabiguero[1]
Larielyn J. Ruazol[1]
and Laiza F. Parago[1]

[1]Department of Electronics Engineering, Technological University of the Philippines, Manila, The Philippines

[2]Center for Engineering Design, Fabrication, and Innovation (CEDFI), College of Engineering, Technological University of the Philippines, Manila, The Philippines

Abstract

The study was conducted in the Payatas Controlled Disposal Facility, which closed permanently in 2017 due to environmental and human health concerns. The area is proposed to undergo redevelopment to promote eco-tourism through ecological park constructions and climate-friendly livelihood projects. Aquaponics is one of the eco-sustainable projects for growing fish and plants. To improve food quality, this study focused on developing a technology-based aquaponic system equipped with multiple sensors, actuators, and Arduino. The essential parameters in fish and plant growth are monitored, such as the pH level, dissolved

* Corresponding Author's Email: edmon_fernandez@tup.edu.ph

In: Aquaponics: Sustainable Farming, Ecology and Innovation
Editor: Lean Karlo Santos Tolentino
ISBN: 979-8-89530-464-8

oxygen and water temperature, light intensity, and air temperature. Moreover, the system utilized the PID algorithm to control the bilge pump to maintain the level of the water flowing to the plants to avoid water overflow. Nile tilapia and romaine lettuce are the types of fish and plants cultured in the system. The performance of the developed aquaponics concluded that there is an upside effect to the tilapias and lettuce in terms of their size and weight, compared to what the traditional aquaponic system produces. The use of technology in the aquaponic system made it more efficient and sustainable, reducing the need for manual monitoring and control. The system's automated sensors and controlled devices enabled control of the environmental factors affecting the growth of fish and plants, providing a promising solution to the challenges faced in sustainable agriculture, particularly in urban areas.

Keywords: aquaponics, PID algorithm, essential parameters monitoring, controlled devices, bilge pump

Introduction

Payatas, a barangay in Quezon City, gained recognition for its dumpsite, nicknamed the "Second Smokey Mountain," due to the continuous disposal of 1,200 tons of garbage daily. The dumpsite's accumulation of waste led to frequent fires and the release of harmful smoke. It has been closed and reopened multiple times (Roxas, 2017), prompted by a tragic incident in July 2000 and concerns over health and environmental impacts. The dumpsite poses health risks and contaminates water resources (Endo, 2017) through the release of leachates and landfill gas. In 2017, it was permanently closed, and plans for its rehabilitation have been under discussion. As part of the rehabilitation efforts, the Department of Science and Technology (DOST) initiated a project to construct an aquaponics system in Payatas. Aquaponics is a sustainable and space-efficient food production system (Rabang, 2019) that has gained popularity in urban areas. Several research studies have focused on automating the monitoring and maintaining water quality parameters in aquaponics setups, such as the study of Shaout and Scott (2017) and Tolentino et al. (2021). Various studies have implemented different sensors and actuators to monitor and maintain parameters in aquaponics. These include an LCD display for water quality parameters, an Android application for remote monitoring and control, and a solar-powered smart aquaponics system with system monitoring and automatic correction.

However, no study has deployed a fully automated smart aquaponics system that focuses on monitoring the growth of plants and fish in the system.

The general objective of this study is to develop a self-sustainable system that produces high-quality food while conserving water, energy, and nutrients by automatically monitoring and maintaining the required parameters. The specific goal is to design and build a fully automated aquaponics setup, incorporating sensors and controlling devices and configuring the necessary parameters using Arduino Mega. The study aims to aid Payatas rehabilitation by implementing a fully automated aquaponics system as part of the DOST project in collaboration with TUP-Manila. Such a system would help Payatas-based farmers save resources, leading to a more efficient method of cultivating fish and vegetables.

Related Studies

Research by Haryanto et al. (2019) explored smart aquaponics as a bio-integrated farming system combined with IoT-based electronic technology. This innovative approach utilizes aquaculture ponds or water containers with feed nutrients as a source of nutrition or hydroponic growing medium. The study findings indicate that the sensors employed, particularly ultrasonic, pH, and temperature sensors, demonstrated high accuracy.

The research by Beecher (n.d.) stated the essential hardware aquaponics components for desired output, such as fish tanks, filtration, NFTs, and aeration. Emphasizing fish tanks must be comfortable and stress-free for the system to function properly. In addition, Pattillo (2017) also stated that plumbing materials are a must. Water movement within the system components is facilitated through plumbing, which can vary in size and material depending on its intended purpose. PVC (polyvinyl chloride) is a popular choice due to its durability, lightweight nature, affordability, ease of handling, and widespread availability, making it highly prevalent in recirculating aquaculture systems. A sump tank is also a good component to consider as it can be the system's central collection point for water, as it operates on gravity. Positioned at the system's lowest point, the sump tank also functions as a reservoir, typically devoid of plants or fish.

Lastly, the study by Azemi et al. (2019) examined and proposed the IoT-based control and monitoring system for aquaponics as it has different sensors that function. The system allows for monitoring water, temperature, and pH levels, which can be accessed through smartphones and browsers.

Additionally, it enables the control of lights and water pumps via the Internet. The controlled environment greatly affects the quality of food production (Ahmed et al., 2020), lettuces, and cultured fish. The system incorporates a Raspberry Pi device as the gateway for sensor readings, an Arduino Uno for monitoring, and an Arduino Nano for control.

Methodology

This developmental research focuses on creating an aquaponics setup that leverages technology to enhance convenience and productivity. The study aims to explore the advantages brought about by these advancements in aquaponics systems. and automated correcting devices were used to monitor water in the fish tank and the plants' environment. This will help provide a conducive environment, which is crucial in culturing fish and plants.

Hardware Development

Figure 1 shows the design of the Aquaponics setup. The fish tank is made of fiberglass and has dimensions of two meters in width, one meter in length, and 0.5 meters in height. Filtering is important to ensure the water is not too acidic and harmful to fish and plants.

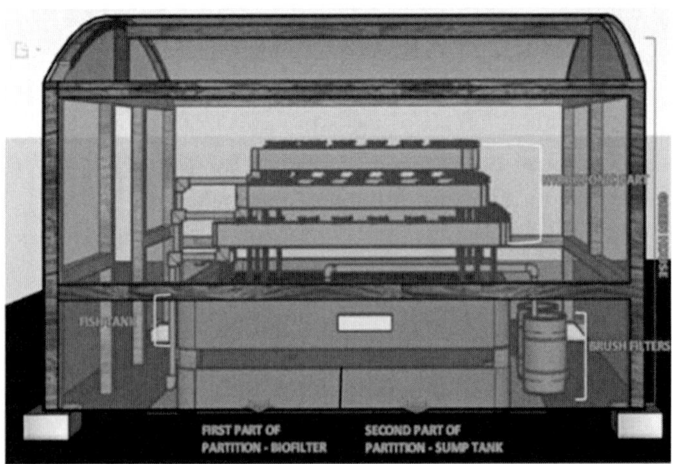

Figure 1. Aquaponics setup design.

Figure 2. Actual image of the aquaponics setup.

Moreover, filters guard against potential pump clogging caused by sediment accumulation in the system. Considering the tank's size, several brush filters are used with a length dimension of 60 cm or 24 inches. Below the fish tank is a sump tank with a dimension of 0.3 meters in height and one meter in length. The water there is meant to be utilized to adjust the pH level of the tank using a water pump. The actual image of the system is shown in Figure 2.

Automation of the Aquaponics System

The block diagram in Figure 3 shows the automation in aquaponics system. The process is divided into two main components: connecting sensors, correcting devices, and integrating an IoT (Internet of Things) system using an Arduino Board. The connections were placed in the chassis to protect it from any contact damage. The actual image of the connections is shown in Figure 4. The PID (Proportional-Integral-Derivative) control algorithm was implemented to control bilge pumps with the aim of maintaining a decent water level flowing through the downspout. Its block diagram is shown in Figure 5. An actual image of its connections is shown in Figure 6.

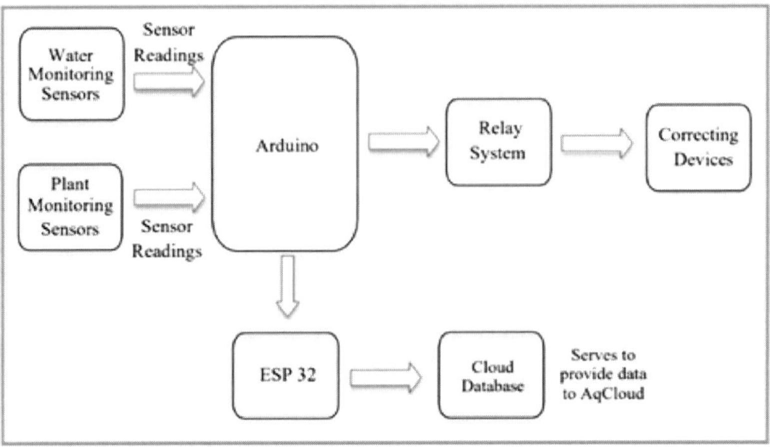

Figure 3. Block diagram.

Figure 4. Wiring connection inside chassis.

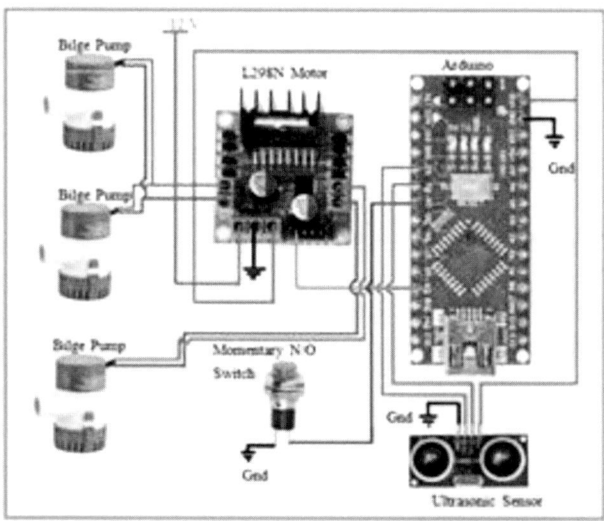

Figure 5. Block diagram with implementation of PID algorithm for water pump control.

Figure 6. Actual image of PID implementation.

Effectiveness Evaluation of the Developed System

A comparison of the sizes of Nile tilapia and romaine lettuces in the SMART Aquaponics with the traditional setup with the same deployment environment was conducted to determine the effectiveness of the developed project. Using a ruler is one of the most convenient tools for measuring something. Thus, using a ruler, we measured the length of the Nile tilapia and the height and length of the lettuce. The culturing of tilapia and lettuce was only within 1 month. Consequently, measuring the byproduct of the SMART and traditional aquaponics was conducted every day to observe a significant change in size, particularly in the fish, where growth takes 3 times longer than the lettuce. Measuring sizes of the tilapia and lettuce are shown in Figures 7 and 8.

Figure 7. Measuring Nile tilapia.

Figure 8. Measuring romaine lettuces.

The sizes gathered from the tilapia and lettuce in one month of culturing were analyzed using a T-test. It aims to determine if there is a statistically significant difference in the means of these variables between the two systems. The T-test formula is as follows:

$$t = (\bar{x}1 - \bar{x}2) / \sqrt{((s12/n1) + (s22/n2))} \quad (1)$$

Where:

$\bar{x}1$ and $\bar{x}2$ are the means of the two samples.
s12 and s22 are the variances of the two samples.
n1 and n2 are the sample sizes of the two samples.

Results and Discussion

The Smart Aquaponics system, incorporating IoT technology, offers an advanced approach to cultivating Nile Tilapia and Romaine Lettuces. By utilizing real-time monitoring and control, this system enables precise adjustments to create optimal growth conditions. Compared to the Traditional Aquaponics system, the Smart Aquaponics system demonstrates superior performance in the growth rates of lettuce height, width, and tilapia length.

An independent one-tail t-test with a statistical significance level of 0.05 was performed for both the Smart and Traditional Aquaponics Setups. Table 1 shows the t-test results comparing the height growth of lettuce in the Smart Aquaponics and Traditional Aquaponics setups. The Smart Aquaponics system achieved an average height growth rate of 50.26%, significantly higher (71.53%) than the 29.3% growth rate in the Traditional Aquaponics system. The null hypothesis (H, assuming no significant difference in growth rates, is rejected, indicating a significant difference in height growth rates between the two systems ($p < 0.05$).

Table 2 shows the t-test results comparing the width growth of lettuce in Smart Aquaponics and Traditional Aquaponics setups. The Smart Aquaponics system achieved an average width growth rate of 54.40%, significantly higher (48.94%) than the 36.52% growth rate in the Traditional Aquaponics system. The null hypothesis (H_0), assuming no significant difference, is rejected ($p < 0.05$), providing evidence of a significant difference in width growth rates between the two systems.

Table 1. T-Test for the height growth rate of lettuce

	Smart Aquaponic Setup	Traditional Aquaponics Setup
Mean	50.26	29.30
Variance	142.21	38.91
Sample Size	12	12
Degree of Freedom	22	
Pooled Variance	90.56	
Hypothesized Mean Difference	0	
t-value	5.396228	
P(T<=t) one-tail	0.000019	
P(T<=t) two-tail	0.000038	

Table 2. T-Test for the width growth rate of lettuce

	Smart Aquaponic Setup	Traditional Aquaponics Setup
Mean	54.40	36.52
Variance	272.64	53.65
Sample Size	12	12
Degree of Freedom	22	
Pooled Variance	163.14	
Hypothesized Mean Difference	0	
t-value	3.427847	
P(T<=t) one-tail	0.001743	
P(T<=t) two-tail	0.003487	

Table 3. T-Test for the length growth rate of lettuce

	Smart Aquaponic Setup	Traditional Aquaponics Setup
Mean	32.66	9.26
Variance	21.42	0.90
Sample Size	25	25
Degree of Freedom	48	
Pooled Variance	11.16	
Hypothesized Mean Difference	0	
t-value	24.768004	
P(T<=t) one-tail	0.00001	
P(T<=t) two-tail	0.00001	

Table 3 shows the t-test results comparing the length growth of tilapia in the Smart Aquaponics and Traditional Aquaponics setups. The Smart Aquaponics system exhibits an average length growth rate of 32.66%, 248.38% higher than the width growth rate of 9.26% in the Traditional Aquaponics system. The null hypothesis (H_0), assuming no significant difference, is rejected ($p < 0.05$), providing evidence of a significant difference in the length growth rates between the two systems.

Conclusion

The Smart Aquaponics system showcased significant capabilities and demonstrated several advantages over traditional setups. The t-test results comparing the growth rates between the Smart Aquaponics and Traditional Aquaponics setups provide statistical evidence of the system's superior performance. Lettuce's average height growth rate in the Smart Aquaponics system was 50.26%, 71.53% higher than the height growth rate of 29.3% observed in the Traditional Aquaponics system (p-value < 0.05). Similarly, lettuce's average width growth rate in the Smart Aquaponics system was 50.26%, 71.53% higher than the width growth rate of 29.3% in the Traditional Aquaponics system (p-value < 0.05). Additionally, tilapia's average length growth rate in the Smart Aquaponics system was 32.66%, 248.38% higher than the width growth rate of 9.26% in the Traditional Aquaponics system (p-value < 0.05).

Furthermore, the survey conducted among experts and owners of smart aquaponics systems provided valuable insights into user satisfaction. The survey results indicated high satisfaction levels across various aspects of the Smart Aquaponics system. Specifically, 60% of the participants agreed that the system met their expectations and requirements, 80% agreed that the interface and controls were intuitive and easy to use, and 100% strongly agreed that the system effectively monitored and controlled water quality, temperature, pH, and nutrient levels. Additionally, 80% of the participants strongly agreed that the system optimally utilized available space to maximize efficiency, and all participants (100%) strongly agreed that the system operated reliably without major technical issues or malfunctions. Moreover, 80% of the participants agreed that the quality of products generated by the system was satisfactory, and they expressed general agreement regarding the number of products.

Based on the statistical data and survey results, further development and optimization of the Smart Aquaponics system for scalability and large-scale production are recommended. This involves increasing capacity, incorporating additional sensors, and focusing on water management techniques. Education and training programs should be established to promote adoption and ensure proper operation. Collaboration among stakeholders is crucial for knowledge sharing and continuous improvement. Future research should explore alternative designs, fish species, and plant varieties. Consideration should be given to using the DYP-A02YY Waterproof Ultrasonic Ranging Sensor for enhanced performance. These actions will optimize the system, contribute to sustainable food production, and address the demand for locally grown produce.

The successful implementation of the Smart Aquaponics System showcases its potential to transform urban agriculture, ensure food security, and promote sustainable practices. Leveraging advanced technologies and exploring innovative approaches will allow the system to evolve and continually deliver even greater future benefits. By addressing limitations, optimizing capabilities, and fostering collaboration, we can enhance efficiency, scalability, and environmental stewardship. It is crucial to continue research and development, invest in education and training, and encourage stakeholder engagement to realize the full potential of smart aquaponics for sustainable food production.

Disclaimer

None.

References

Ahmed, H. A., Yu-Xin, T., & Qi-Chang, Y. (2020). Optimal control of environmental conditions affecting lettuce plant growth in a controlled environment with artificial lighting: A review. *South African Journal of Botany,* 130, 75–89. doi:10.1016/j.sajb.2019.10.020.

Azemi, N. C., et al. (2019). IoT-based intelligent greenhouses (IGH) using Lo-Ra technology. *International Journal of Innovation, Creativity and Change,* 9(11), 274-283. Retrieved from www.ijicc.net.

Beecher, L. (n.d.). *Aquaponics: System Layout and Components* (no. figure 3).

Endo, J. (2017). The Mountain of Garbage that Blights the Philippine Capital. Nikkei Asia. Retrieved from https://asia.nikkei.com/NAR/Articles/The-mountain-of-garbage-that-blights-the-Philippine-capital.

Haryanto, M., Ulum, M., Ibadillah, A. F., Alfita, R., Aji, K., & Rizkyandi, R. (2019). Smart aquaponic system based on the Internet of Things (IoT). In *Journal of Physics: Conference Series* (Vol. 1211, No. 1). doi:10.1088/1742-6596/1211/1/012047.

Pattillo, D. A. (2017). An overview of aquaponic systems: Aquaculture components. *NCRAC Technical Bulletin,* 20. Retrieved from http://lib.dr.iastate.edu/ncrac_techbulletins/20.

Rabang, I. (2019). Indoor Farming: A Snapshot of Sustainable Aquaponics System. Retrieved from https://www.boldbusiness.com/infrastructure/aquaponics-system aquarium/.

Roxas, P. A. (2017, August 6). Environmentalists hail closure of Payatas dumpsite. INQUIRER.net. *Quezon City.* Retrieved from https://newsinfo.inquirer.net/920737/environmentalists-hail-closure-of-payatas-dumpsite.

Shaout, A., & Scott, S. G. (2017). IoT fuzzy logic aquaponics monitoring and control hardware real-time system. In *Proceedings of the International Arab Conference on Information Technology.* Retrieved from https://acit2k.org/ACIT/images/stories/year2014/month1/ACIT2017_Proceeding/139.pdf.

Tolentino, L. K. S., et al. (2021). Development of an IoT-based Intensive Aquaculture Monitoring System with Automatic Water Correction. *International Journal of Computer and Digital Systems,* 10(1), 1355–1365. doi: 10.12785/ijcds/1001120.

Index

F

G

H

I

K

L

M

N

O

P